MORT SCHULTZ'S
Electronic
Fuel Injection
Repair Manual

Also by Mort Schultz

Crown's Diesel Repair Manual
Your Ford
Get Your Chevrolet Fixed Right
Get Your Pontiac Fixed Right
Get Your Buick Fixed Right
Get Your Oldsmobile Fixed Right
How to Install Dealer Options in Your Own Car
1001 Questions and Answers About Your Car
The Popular Mechanics Complete Car Repair Manual
The McGraw-Hill Illustrated Home Study Automotive Course
How to Fix It
The Practical Handbook of Painting and Wallpapering
The Popular Mechanics Complete Appliance Repair Manual
Wiring
Teaching Ideas That Make Learning Fun
The Teacher and Overhead Projection
Photographic Reproduction

Also, contributing editor/writer for the following:

The Reader's Digest Complete Car Repair Manual
Time-Life Book of the Family Car
Motor Import Car Repair Manual
Reader's Digest Home Improvements Manual
Reader's Digest Book of Home Projects
Popular Mechanics Do-It-Yourself Encyclopedia
America's Handyman Book

MORT SCHULTZ'S Electronic Fuel Injection Repair Manual

Troubleshooting and Repairing Electronic Fuel Injection Systems

General Motors • Ford • Chrysler • Foreign Cars

Mort Schultz

Crown Publishers, Inc.
New York

Grateful acknowledgment is hereby given to the following companies for permission to reprint excerpted materials from their publications:

Artwork from the General Motors Product Service Training publication entitled "Fuel Injection" (#16009.10-1A) courtesy of the General Motors Corporation

Illustrations from *Electronic Fuel Injection* copyright © 1982 by Nissan Motor Corporation, courtesy of Nissan Motor Corporation in the U.S.A.

Artwork from Ford Motor Parts and Service Division publication NO. 2670-001, September 1984, entitled "Technician's Reference Book—Electronic Fuel Injection" courtesy of the Ford Motor Corporation

Illustrations from Chrysler publications courtesy of Chrysler Motors

Copyright © 1990 by Mort Schultz

All rights reserved. No part of this book may be reproduced or transmitted in any form or by any means, electronic or mechanical, including photocopying, recording, or by any information storage and retrieval system, without permission in writing from the publisher.

Published by Crown Publishers, Inc., 201 East 50th Street, New York, New York 10022

CROWN is a trademark of Crown Publishers, Inc.

Book design by J. Victor Thomas

Printed in the U.S.A.

Library of Congress Cataloging-in-Publication Data

Schultz, Morton J.
  Mort Schultz's electronic fuel injection repair manual/by Mort Schultz.
  Includes index.
  1. Automobiles—Motors—Electronic fuel injection systems—Maintenance and repair—Handbooks, manuals, etc. I. Mort Schultz. II. Title. III. Title: Electronic fuel injection repair manual.
TL214.F8S38   1989   89-1270
629.2'53—dc19
ISBN   0-517-57240-0
10 9 8 7 6 5 4 3 2 1
First Edition

Flexibook

Contents

	Introduction	vii
1	**Electronic Fuel Injection** *The Role of Vacuum and Electricity*	1
2	**GM Throttle Body Fuel Injection Systems** *An Overview*	11
3	**GM Throttle Body Fuel Injection Systems** *How They Work*	17
4	**GM Throttle Body Fuel Injection Systems** *Troubleshooting and Repairing Fuel Delivery Components*	23
5	**GM Throttle Body Fuel Injection Systems** *Servicing the Throttle Body*	33
6	**GM Multiport Fuel Injection System** *Preliminary Troubleshooting and Repair*	43
7	**GM Multiport Fuel Injection System** *Troubleshooting and Repairing Other Components*	51
8	**Ford Central Fuel Injection Systems** *An Overview*	61
9	**Ford Central Fuel Injection Systems** *Troubleshooting and Repair*	71
10	**Ford Multipoint Fuel Injection Systems** *Troubleshooting and Repair*	85
11	**Chrysler Single-Point Electronic Fuel Injection System** *Electronic Considerations*	99
12	**Chrysler Single-Point Electronic Fuel Injection System** *Makeup of the System*	105

| 13 | **Chrysler Single-Point Electronic Fuel Injection System** | 111 |
| | *Troubleshooting and Repair* | |

| 14 | **Chrysler Multiport Fuel Injection System** | 123 |
| | *An Overview* | |

| 15 | **Chrysler Multiport Fuel Injection System** | 133 |
| | *Troubleshooting and Repair* | |

| 16 | **Foreign Car Electronic Fuel Injection Systems** | 141 |
| | *How They Work* | |

| 17 | **Foreign Car Electronic Fuel Injection Systems** | 149 |
| | *Preliminary Troubleshooting and Repair* | |

| 18 | **Foreign Car Electronic Fuel Injection Systems** | 155 |
| | *A Part-by-Part Check* | |

Index   164

Introduction

IF YOUR PRESENT CAR doesn't have an electronic fuel injection (EFI) system, one of your future cars most certainly will. As this is being written, EFI systems are being installed in approximately 95 percent of the new vehicles coming off the production line. By 1990, that number will be almost, if not totally, 100 percent.

Mort Schultz's Electronic Fuel Injection Repair Manual, therefore, is written for every car owner and professional mechanic who is having trouble making sense out of EFI. One of the goals of this book is to prove to you that EFI is the easiest to understand and easiest to work on automotive fuel system in existence—easier and much simpler in makeup, performance, troubleshooting, and repair than carburetor fuel systems, which have been with us since the inception of the automobile more than one hundred years ago. If you have ever worked on a carburetor fuel system (even if you haven't), give EFI a chance. You'll eventually find it a breeze to troubleshoot and repair.

There is a lot of misunderstanding concerning EFI. Much of it has been generated by the use of different terms that mean the same thing. This leads one to believe that there are numerous types of fuel injection systems when, in fact, there are only two: throttle body fuel injection (TBI) and multiport fuel injection (MPFI). MPFI is also called multipoint fuel injection. Assuredly, there are variations of these two from one manufacturer to another. This book addresses these variations and also attempts to standardize the various terms.

The author wishes to thank the following companies for providing technical illustrations and support: Chrysler Motors, Ford Motor Company, General Motors Corporation, Robert Bosch of Germany, and Nissan Motor Corporation in the U.S.A.

1

Electronic Fuel Injection

The Role of Vacuum and Electricity

AUTOMOBILE MANUFACTURERS AGREE that most performance problems with cars having electronic fuel injection (EFI) systems are caused by vacuum and electrical failures—not by a malfunction with the EFI. Therefore, being aware of possible vacuum and electrical malfunctions is to your advantage. You may be able to resolve a problem by using a procedure in this chapter, thus avoiding having to tackle the fuel injection system.

FACTS ABOUT VACUUM

What is referred to by most mechanics as engine vacuum is not vacuum at all: It is negative pressure. Vacuum is space that is void of everything, including air. The "vacuum" created in an intake manifold when the engine is running, however, is pressure that is less than whatever the atmospheric pressure happens to be in the manifold before the engine is started.

Thus, if atmospheric pressure in the intake manifold is 14.7 pounds per square inch absolute (psi) when the engine is not running and 11.5 psi when it is running, the 11.5 psi is negative pressure, or vacuum.

Whether you call it vacuum or negative pressure, it is essential to check it when trying to track down a performance problem with an engine that is equipped with an EFI system. There are two reasons for this check: (1) to see that there is no cutoff of negative pressure to components that need it to function and (2) to be sure no component that is getting the negative pressure supply it needs is damaged and malfunctioning in any way.

VACUUM-RELATED PROBLEMS

Depending upon which components are affected, the problems a disruption in negative pressure causes are stalling, hard starting, hesitation (also known as sag and stumble when accelerating), too fast an idle, rough idling, surging, lack of power, missing, backfire, poor fuel economy, dieseling (also known as engine run-on when the ignition is turned off), pinging (also known as detonation, autoignition, and spark knock), failure to pass a state emissions test, and a hydrogen sulfide (rotten egg) odor from the exhaust.

However, a malfunction in a fuel injection system can also cause one or more of these headaches, so why should you begin troubleshooting with vacuum and not with EFI? Because it is estimated that vacuum leaks and damaged vacuum components account for at least 60 percent of engine drivability problems, while problems brought on by faulty EFI components account for 25 percent—that's why. The remaining 15 percent are electrical-related.

Components that need vacuum to operate are found all over cars equipped with EFI (Fig. 1-1). Depending on the engine and the type of fuel injection system involved, the vacuum components include the exhaust gas recirculation (EGR) valve, positive crankcase ventilation (PCV) valve, distributor vacuum advance, thermostatically controlled air cleaner vacuum motor, deceleration valve, throttle modulator, manifold absolute pressure (MAP) sensor, charcoal canister purge valve, air bypass valve, distributor modulator valve assembly, exhaust heat control valve, and a variety of vacuum switches, vacuum control valves, and vacuum regulators.

Nonengine parts also use vacuum created by the engine. These include the brake booster, automatic transmission, speed (cruise) control, and climate control system.

Negative pressure created inside the engine by the action of the pistons "gets" to parts through hoses connected to fittings (vacuum taps) on the EFI throttle body and/or

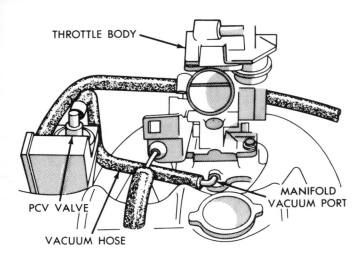

Fig. 1-1. The location of vacuum components differs from engine to engine. Here you are looking at a positive crankcase ventilation (PCV) valve on a late-model Chrysler engine. To service your engine, you will need the diagram that shows the vacuum hose routing. If this diagram isn't mounted in the engine compartment, look for it in the shop manual for the vehicle.

intake manifold. These hoses are the weak links in the system, so it is with them that you should begin to troubleshoot a possible disruption in vacuum.

SOUNDS AND SIGHTS

Don't get scared off by what looks like a jumble of spaghetti under the hood. Although vacuum hoses seem to be everywhere, it is not always hard to pick out the one that is leaking.

A leaking hose usually announces its existence by hissing or whistling. To make it easier to pinpoint which hose among the many you see is causing a vacuum loss, use a 4-foot length of 5/16-inch vacuum hose as a stethoscope. With the engine idling, hold one end of the hose to your ear and move the other end slowly over each vacuum hose (Fig. 1-2). Be sure to scan the component and vacuum tap to which the hose connects: The leak might be at a connecting point.

If a vacuum hose is leaking where it connects to a vacuum tap on the engine or to a fitting of the component it serves, secure it and test it again (Fig. 1-3). If the hissing or whistling has stopped, you're home free. But if the hissing or whistling continues, the hose is cracked at that point or the part is damaged and losing vacuum. The defective hose or component should be replaced.

Keep in mind that a hose doesn't have to be leaking for it to be the cause of a negative pressure loss. It could be bent or blocked with debris. This possibility is more likely when an engine performance problem develops immediately after you have done some work on the engine. You may have kinked or closed off a hose by accidentally shoving a part, such as the air cleaner, down on it.

The chance of a hose getting clogged with debris is remote, but squeeze each hose from

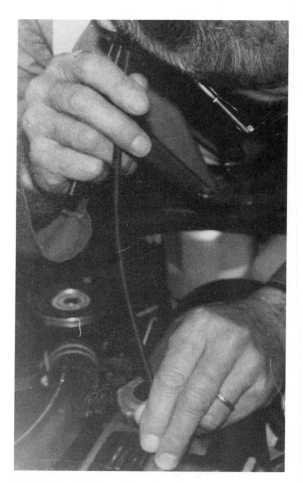

Fig. 1-2. A 4-foot length of 5/16-inch vacuum hose makes a handy stethoscope to uncover a damaged on-the-car vacuum hose or loose vacuum connection that is leaking.

end to end anyway to make sure it hasn't happened. If there is something in there, you will feel it. Remove the hose and blow out the debris.

CHECK THE CHART

Use the vacuum hose routing chart mounted in the engine compartment to identify vacuum hoses and vacuum components. If this decal is missing or has been obliterated, you will have to trace each hose back from its fitting on the throttle body and intake manifold to find branch hoses and vacuum components that intersect with it. If the car has a throttle body system, this procedure entails first removing the air cleaner to locate vacuum taps on the throttle body and intake manifold. If the car has a multipoint EFI system, look for vacuum taps on the manifold.

To make future troubleshooting easier, put a dab of the same color paint on each hose and the component it serves: a dab of blue, for example, on the canister purge valve and its hose, and a dab of red on the EGR valve and its hose (Fig. 1-4).

MORE SOPHISTICATED TROUBLESHOOTING

When the sensory methods (listening, looking, and feeling) fail to reveal a disruption of

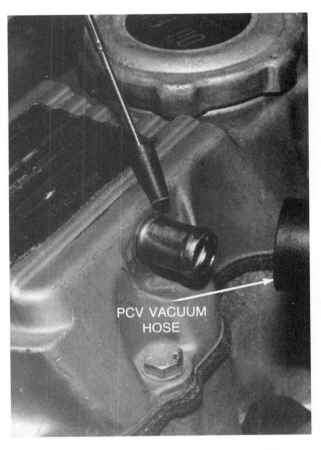

Fig. 1-3. Secure vacuum hoses are necessary for sound engine performance. Make sure, for instance, the hose that delivers vacuum to the PCV valve, shown here without the hose connected to it, is tight.

Fig. 1-4. Once you've identified the vacuum hose that serves a component, place a mark on both the hose and the component to make identifying easy in the future. This EGR valve has been marked with a black dot and so has its hose, which is hidden from view.

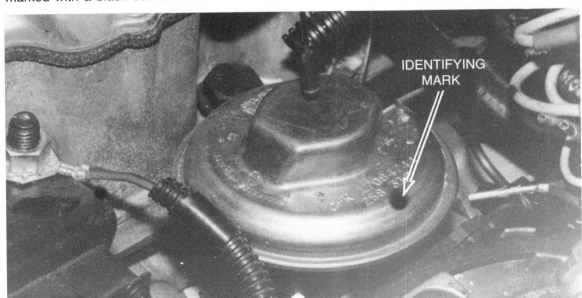

vacuum, it's time to use a digital volt/ohmeter (VOM) having an impedance of 10 mega-ohms. Locate the oxygen (O_2) sensor and disconnect it from the electrical feed wire. The O_2 sensor is screwed into the exhaust manifold to monitor the oxygen content of the exhaust. It is connected to the computer.

An O_2 sensor has either a single terminal or two terminals. O_2 sensors with single terminals are grounded internally. O_2 sensors with two terminals have one wire, the feed wire, extending to the computer and the second wire bolted to the engine. This is the ground.

If the O_2 sensor has a single terminal, attach the alligator clips of two jumper wires to the metal (probe) end of the VOM positive lead. Then connect the alligator clip on the other end of one jumper to the terminal of the O_2 sensor. (See Chapter 2, Fig. 2-9.) Shove the clip on the end of the other jumper into the plug of the O_2 sensor electrical feed wire, making sure there is a good metal-to-metal contact. Finally, attach the VOM negative lead to ground.

If the O_2 sensor has two terminals, connect jumpers as described above, but also connect a third jumper wire between this second terminal and ground. That second terminal is a ground terminal. If you don't get a proper response when making the test, the connections to the two O_2 sensor terminals are mixed up, so reverse the jumpers.

After connecting the VOM, set the instrument to the 2-volt DC scale, start the engine, and let it run at idle. If the VOM needle swoops sharply from low (nearly 0, in fact) to high (in the 0.6 to 0.8 range) and back, there is no vacuum leak. Neither is there a vacuum leak if the VOM reading stays in the high range, but you do have a rich-running condition on your hands. Test the EFI system followed, if necessary, by the computer control system.

If the VOM reading stays in the low range, which indicates a lean-running condition, there very well could be a disruption of vacuum. A hand-held vacuum pump will help you find the area where a vacuum loss may exist.

USING A VACUUM PUMP

The VOM O_2 sensor troubleshooting method will give you an indication of a possible vacuum loss. But when push comes to shove, a vacuum gauge is a more accurate instrument.

If you are in the market for one, consider a hand-held vacuum pump that incorporates a gauge (Figs. 1-5a, 1-5b). The instrument serves more than one purpose:

■ As a gauge it can help you determine if engine vacuum is to specification and also if there is a non-vacuum-related defect.

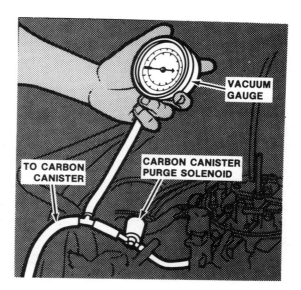

Fig. 1-5a. There are two types of instruments available to test engine vacuum. The one shown here is a vacuum gauge. By connecting it to a vacuum source, in this case the vacuum line between the carbon canister of a fuel evaporation control system and a purge solenoid, you can determine if there is a vacuum deficiency somewhere in that circuit. The vacuum gauge will not, however, point out which component is damaged.

Fig. 1-5b. The hand-held vacuum pump is the other instrument that detects engine vacuum. It also reveals a problem in a vacuum circuit. You can then attach the pump to each component to determine which one is defective. Figure 1-6 demonstrates how this is done.

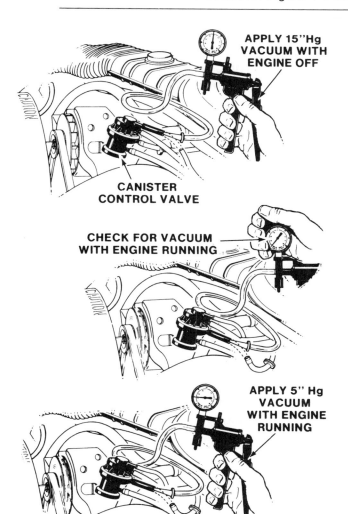

- As a detection instrument it can help you find a leaking vacuum hose or a damaged vacuum component (Fig. 1-6).

As a Vacuum Gauge

Vacuum gauges are calibrated in terms of inches of mercury (in. Hg) instead of atmospheric pressure to make them easier to interpret. They are set to give a reading of 0 in. Hg when the engine isn't running instead of a reading that designates atmospheric pressure when the test is being made.

You need to know the exact vacuum

Fig. 1-6. A hand-held vacuum pump allows thorough testing of components. In testing a fuel evaporation system canister control valve, for example, you first pump up vacuum with the engine off and see if it holds for at least 20 seconds. If not, replace the valve. You then check the valve with the engine running and with the system in various modes. In one mode, there should be no indication of vacuum. In another, there should be. You need the shop manual, which outlines how to make the various tests.

specification for your engine. The range of 17 to 21 in. Hg that is recommended by many general automotive repair manuals is not applicable to most modern engines. Using it can lead to false test conclusions.

For example, a Buick 3.8-liter V-6 engine with overlapping valves has a normal vacuum of 15 in. Hg. "Overlapping" means that during a combustion cycle the exhaust and intake valves in a cylinder are both open simultaneously for a brief period. If you use that 17 to 21 in. Hg range on this engine, you may go in search of a vacuum leak that doesn't exist.

In checking negative pressure with a vacuum gauge, keep in mind that specifications in service manuals are sea level readings. All readings will drop by 1 in. Hg for every 1,000 feet you are above sea level. For example, a reading of 20 in. Hg in New York City (sea level) will be 15 in. Hg in Denver (which is 5,000 feet above sea level) for the same engine. Therefore, if you are in Denver and you don't adjust the service manual reading, you will go wrong.

To see if the vacuum developed by an engine is up to snuff, you have a few choices. You can disconnect a vacuum hose from a vacuum tap on the throttle body or intake manifold and connect the gauge to the fitting. Or you can disconnect the vacuum hose from the brake booster, remove the check valve if that part is pressed into the end of the hose, and connect the gauge to the hose. You will probably have to use an adapter to make the connections. Or you can disconnect a vacuum hose from an engine vacuum component and connect the gauge to the end of the hose. Again, you'll probably have to use an adapter.

However you choose to do it, the connection must be secure. If there is any leak between the gauge and the source of vacuum, you won't get a true vacuum reading.

There's another important point here: If the engine has a throttle body, make sure the source of the vacuum into which you tap the gauge is manifold vacuum—not ported vacuum. Manifold vacuum is true engine vacuum. Ported vacuum is less than manifold (true) vacuum and will therefore lead to an erroneous test reading. A manifold vacuum tap is positioned below the throttle plate of the throttle body. A ported vacuum source lies above the throttle plate.

And there is still another precaution: If you decide to attach your gauge to a main vacuum hose, see that the hose doesn't possess a vacuum restrictor upstream from where the gauge is connected. This will affect the reading.

Start the engine and allow it to warm up. With the engine running at idling speed, the vacuum gauge should show a normal reading that holds steady. The gauge needle should neither waver nor drift.

Then, have an assistant in the car open and close the throttle rapidly by pressing and releasing the accelerator pedal. The gauge needle should drop below 5 in. Hg, bounce up to 2 or 3 in. Hg above normal, and then settle at normal.

If the needle holds steady at idling speed, but in the 4 to 7 in. Hg range, this indicates a loss of vacuum from the engine itself. Most often this loss is caused by loose throttle body mounting bolts or a bad gasket between the throttle body and intake manifold. Otherwise, piston rings may be worn or there may be a loss of negative pressure around the valves. Do dry and wet compression tests to check out these possibilities.

One way to check the integrity of the throttle body-to-manifold joint is to make a mixture of kerosene and some SAE 20 or 30 motor oil and spread it around the seam as you keep your eye on the vacuum gauge. If the gauge needle rises or engine idling speed changes its pitch even momentarily, there is a leak. If tightening the bolts doesn't work, remove the throttle body and replace the gasket.

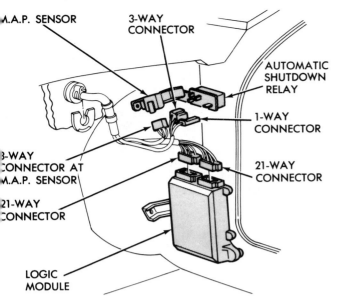

Fig. 1-7. As this drawing illustrates, there are many electrical connectors in a car. Finding one that is corroded or loose is difficult without an electrical circuit schematic drawing. There is one for your vehicle in the service manual.

Interpreting Non-Vacuum-Related Defects

Suppose the vacuum gauge needle doesn't show a normal reading, but neither does it show a loss of negative pressure. This is what you may be up against:

A steady reading of 8 to 14 in. Hg at idling speed: Ignition timing isn't set properly or there is a compression leak past the piston rings. A vacuum gauge needle that wavers slowly by 2 to 6 in. Hg from normal also suggests a compression leak.

A sharp back-and-forth fluctuation of about 5 in. Hg from normal at idling speed: Suspect a leaking head gasket or worn valve guides.

A steady reading above normal at idling speed: There is a restriction in the air intake system. Look for a clogged air filter.

A reading that drops to near zero and rises but doesn't hit normal when the throttle is opened and closed rapidly: Look for a restriction in the exhaust system.

As a Detection Device

Along with cracked hoses and a bad throttle body gasket, another cause of engine problems because of vacuum disruption is a split vacuum component diaphragm. Every vacuum component in a car has a rubber diaphragm inside it that is necessary for the retention of vacuum. If this diaphragm splits, it would prevent a vacuum component from functioning although it is getting vacuum.

You need patience to find out if there is a bad diaphragm. It's a matter of testing one component after the other until you hit the one you are looking for. Disconnect the hose from the vacuum fitting of each component, attach the hand vacuum pump to the fitting and apply vacuum. If the gauge needle doesn't rise or if it rises and falls, the diaphragm is defective. Replace the component.

FAULTY ELECTRICAL CONNECTORS

There are many plastic-molded electrical connectors on an engine equipped with EFI. Any one of them may be loose or corroded (Fig. 1-7). This would cause a disruption of electric current to key parts of the engine, including fuel injectors, and would make the engine behave poorly (Fig. 1-8).

As with vacuum components, troubleshooting electrical connectors is often a hit-and-

Fig. 1-8. A simple way to determine if poor engine performance is caused by a loose electrical connector is to let the engine run as you wiggle various connectors. If you hear a change in the way the engine runs, you've uncovered a faulty connection.

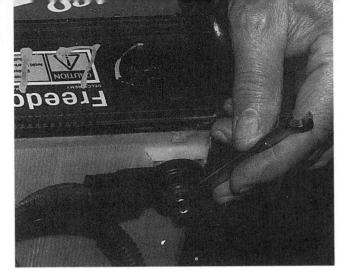

Fig. 1-9. Before working on electrical connectors, disconnect the negative (ground) cable from the battery. When you reconnect that cable, make sure it is secure.

Fig. 1-11. Determine if grease had been applied to the connector pins. If not, do not treat with grease. Leave the connector dry.

miss proposition to find the connector that is corroded. To service each connector, do the following:

1. Remove the negative cable from the battery (Fig. 1-9).

2. Grasp the two parts of the connector, press locking tabs (if there are any), and pull the connector apart (Fig. 1-10).

3. Clean corrosion out of the connector with a small brush. Inspect inside to make sure terminal pins are straight. Use needle-nose pliers to bend those that aren't upright.

4. If there was grease in the connector when you opened it, spread dielectric grease into one half (Figs. 1-11, 1-12). If there was no grease, do not apply any.

Note: You can get dielectric grease from a store that sells electronic parts.

5. Press the halves together firmly. The connector must be secure. You should not be able to pull it apart without a good deal of force.

Fig. 1-10. Connectors are usually locked together with tabs. To separate the two parts, press the tabs and pull.

Fig. 1-12. If a connector had been treated with grease, spread a little dielectric grease in one half of it. Then, secure the two halves of the connector.

2 GM Throttle Body Fuel Injection Systems

An Overview

Fig. 2-1. Is this air cleaner covering a carburetor or a fuel injection throttle body? To find out, remove the air cleaner.

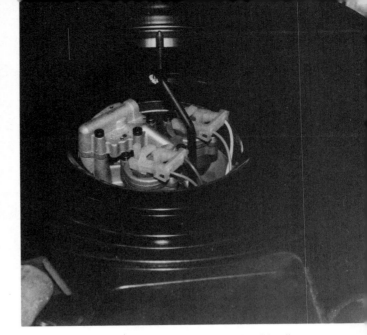

Fig. 2-3. This TBI system is a two-point (dual) unit. Note the two electrical connectors.

ON THE SURFACE, the different varieties of throttle body fuel injection (TBI) systems GM uses look like carburetors. Each is mounted on the engine intake manifold as a carburetor would be, and each has the large familiar-looking air cleaner housing over it, as a carburetor does (Fig. 2-1). Once you remove the air cleaner, however, differences between TBI and a carburetor become apparent.

Immediately, you notice the presence of one or two fuel injectors (Figs. 2-2, 2-3, 2-4). A fuel injector is the component that sprays gasoline into the throttle body.

Fig. 2-2. As with a carburetor air cleaner, a fuel injection system throttle body air cleaner houses an air filter element. Keep this element clean to prevent engine flooding.

Fig. 2-4. The GM single-point TBI system has one fuel injector (arrow). It is served by one electrical connector that is attached to terminals on the injector. It is through this connector that the injector receives current which causes it to open and spray gas into the throttle body.

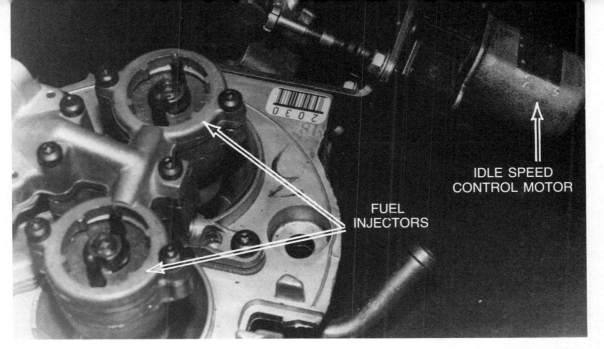

Fig. 2-5. The GM two-point TBI system has two fuel injectors. Each has an electrical connector that attaches to a two-pin electrical terminal on top of the fuel injector. Also seen in this photograph is the idle speed control motor, which is discussed in Chapter 3.

GM's single-point TBI system uses one injector (Fig. 2-4). The two-point (dual) GM TBI system employs two fuel injectors (Fig. 2-5). There is also a third system, called the GM two-by-one (2 × 1) or cross-fire TBI system. It has two throttle body assemblies mounted adjacent to each other. Each throttle body is equipped with one fuel injector (Fig. 2-6).

GM varies its use of these three TBI systems from engine to engine. For this reason, trying to pinpoint which system goes with which engine is not a good way to identify the system.

Fig. 2-6. The GM 2 × 1 (cross-fire) TBI system consists of two single-point assemblies that are coupled together. It is used to serve GM V-8 engines. Each unit feeds gas to the bank of four cylinders on its side of the engine.

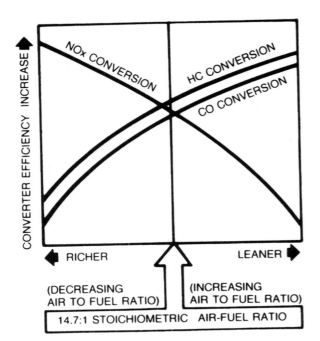

Fig. 2-7. The ideal fuel mixture of 14.7 parts of air to 1 part of gas is called the stoichiometric ratio. The mixture burns completely in an engine and therefore doesn't produce any of the three main polluting agents—carbon monoxide (CO), hydrocarbons (HC), and oxides of nitrogen (NO_x). An EFI system that is working as it should blends air and gas together as close to the stoichiometric value as possible.

The reasonable approach is to become acquainted with each system, so you will be able to identify at a glance the one you are dealing with.

The purpose of a TBI system—single injector, dual injector, or cross-fire—is to help manufacturers meet the stringent exhaust emissions and fuel efficiency standards that have been established by the federal government. This is the main reason why carburetors have been retired in favor of fuel injection systems.

The cost of trying to perfect carburetor-related equipment to meet government exhaust emissions and fuel efficiency standards is high. The carburetor would require extensive modification to meet these standards. Fuel injection does the job satisfactorily and at less expense.

THE STOICHIOMETRIC RATIO

The air and gas mixture that burns most completely in an engine's cylinders is 14.7 parts of air to 1 part of gasoline. When ignited, practically all of that mixture is burned in the cylinders. Thus, the least possible amount of carbon monoxide and hydrocarbons is emitted from the tail pipe to pollute the atmosphere. Furthermore, the buyer of gasoline benefits from every penny's worth, since practically every drop of gas is used to run the engine. This ideal 14.7 to 1 proportion of air to gas is called the *stoichiometric ratio* (Fig. 2-7).

An electronically controlled fuel injection system that is working properly provides a fuel mixture that is closer to the stoichiometric value than any yet developed. Although the feedback (electronic) carburetor, which is one of the expensive carburetor modifications alluded to above, permits a precise amount of gas and air to reach the carburetor, it wasn't until EFI entered the picture that auto manufacturers got the ability to provide a highly efficient fuel delivery system that can control emissions at relatively low cost.

THE ROLE OF THE COMPUTER

GM TBI and multiport fuel injection systems (Chapters 6 and 7) are controlled by a system that consists of a computer and sensors that serve the computer (Fig. 2-8). GM calls this system computer command control (C3).

The C3 and EFI systems are closely related in that the former controls the latter. If the C3 system develops a malfunction, it can cause the EFI system to deliver an unbalanced fuel mixture to the engine. The result will be a disruption of engine performance. Therefore, it is beneficial to have some knowledge of the C3 system. If you do not find the cause of an EFI-suspected drivability problem within the EFI system, in all likelihood the reason for the trouble will be located somewhere in the computerized system.

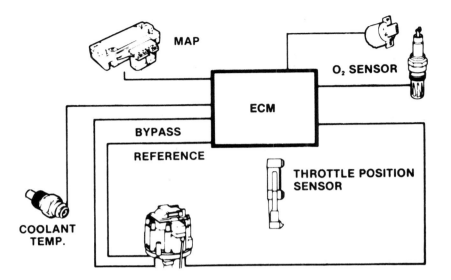

Fig. 2-8. The ECM, or computer, is at the heart of the GM C3 system. Some important sensors that feed data to the ECM are the MAP, O_2 (exhaust oxygen), throttle position, and coolant temperature sensors.

The purpose of the computer is to gather data from a number of sensors (some sensors are referred to as switches) positioned in and on the engine. These devices relay information to the computer concerning engine operation. Incidentally, the computer in GM vehicles is called the electronic control module (ECM) by GM.

Based on data supplied to it by sensors, the ECM determines the quantity of fuel needed by the engine to accommodate the conditions under which it's operating. In other words, the ECM controls the fuel injectors to get those devices to deliver the precise amount of gas necessary for the engine to function at maximum efficiency while cranking and starting, while idling, while accelerating, while cruising, or while decelerating.

No one GM engine equipped with EFI and C3 possesses all of the following sensors, but each has most of them:

- air conditioner on/off switch
- engine coolant temperature sensor
- engine crank sensor
- O_2 sensor (Fig. 2-9)
- crankshaft position sensor
- engine speed sensor
- vehicle speed sensor
- MAP sensor
- park/neutral switch
- system voltage switch
- throttle position sensor
- transmission gear position switch
- power steering pressure sensor
- mass air flow sensor
- manifold air temperature sensor

- EGR vacuum sensor

- engine knock sensor

- barometric pressure sensor

- differential pressure sensor.

In addition to the fuel injection system, the C3 system of a GM vehicle controls the operation of the ignition and emissions control systems as well as the transmission converter clutch, air conditioner, and engine cooling fan.

The key to determining if there is a C3 malfunction lies with the CHECK ENGINE or SERVICE ENGINE SOON light on the instrument panel. A GM model has either a CHECK ENGINE or SERVICE ENGINE SOON light. Both do the same thing. If the light doesn't go off as soon as the engine is started or comes on as you are driving along, it is signaling a malfunction within the C3 system.

You may on occasion see the CHECK ENGINE or SERVICE ENGINE SOON light flicker briefly. As long as the engine is performing normally, there is no cause for concern. Flickering is frequently the result of the C3 system readjusting itself.

There is a chance that a malfunction in the C3 system may not be accompanied by a CHECK ENGINE or SERVICE ENGINE SOON light that stays on. The light may flash briefly, during which time you may not notice it. However, a malfunction in the computerized system will always result in a trouble code being recorded by the computer. Therefore, if troubleshooting the EFI system as discussed in Chapters 1 through 7 fails to reveal a malfunction, turn your attention to the C3 system.

Fig. 2-9. The O_2 sensor is a key unit: It supplies data to the ECM about the oxygen content of the exhaust—whether there is too much (overly rich fuel mixture) or not enough (lean fuel mixture). The ECM can then make adjustments to alter the length of time fuel injectors spray gas into the cylinders.

3

GM Throttle Body Fuel Injection Systems

How They Work

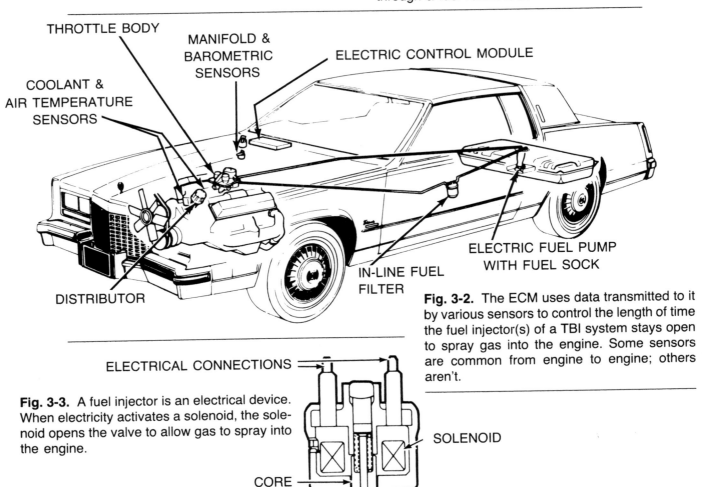

Fig. 3-1. The TBI system is simple. Gas is delivered by the fuel pump from the fuel tank to the throttle body through an in-line fuel filter and fuel delivery line. Excess gas not delivered to the engine is returned to the fuel tank through a fuel return line.

Fig. 3-2. The ECM uses data transmitted to it by various sensors to control the length of time the fuel injector(s) of a TBI system stays open to spray gas into the engine. Some sensors are common from engine to engine; others aren't.

Fig. 3-3. A fuel injector is an electrical device. When electricity activates a solenoid, the solenoid opens the valve to allow gas to spray into the engine.

GAS IS PUMPED from the gas tank into the fuel circuit by an electrically operated fuel pump in the tank. Before gas gets to a fuel injector, which delivers it to the combustion chambers, it passes through a fuel filter (Fig. 3-1).

It's as simple as that. But the important aspect of getting gas to the engine is control: when to deliver it and in what quantity so the engine will run efficiently. The key component doing the control is the ECM (Fig. 3-2).

As the throttle is opened by the driver, the ECM activates a solenoid in the injector, which causes a valve to lift, allowing gas to spray into the throttle body (Fig. 3-3). Gas then flows through a plate in the bottom of the throttle body, called the throttle plate or valve, into the engine intake manifold.

The amount of fuel that flows into the engine depends on how long the ECM allows the injector to stay open. This, in turn, depends on data the ECM gets from the various sensors. Since the injection on-time (or pulse width) varies with the way that the engine is being operated (for example, idling, accelerating, or cruising), it is necessary to have pressure within the fuel system at a constant level at all times. If pressure were to vary, either not enough gas would flow into the engine at any one moment—causing a lean mixture—or too much gas would flow in—causing an overly rich mixture.

Keeping pressure constant in the fuel circuit of a GM TBI system is the job of the fuel pressure regulator (FPR). By maintaining pressure at a constant level, the FPR ensures that the volume of gas reaching the fuel injector is consistent at all times.

As was already mentioned, the amount of gas that's allowed to spray from an injector into the engine intake manifold depends on the length of time the ECM keeps the injector open. As the vehicle's operational conditions change, however, to operate efficiently the engine may need less gas than that reaching the injector. What happens to the gas that reaches the injector that is not needed? It is diverted back into the gas tank through a hose called the fuel return line, which extends from the FPR to the gas tank (Fig. 3-4). Some gas always flows back to the tank through the fuel return line, but under some operating conditions more gas is returned than at other times.

AIR FORCE

Gasoline by itself won't burn. It needs air. With GM TBI systems, this air enters the throttle

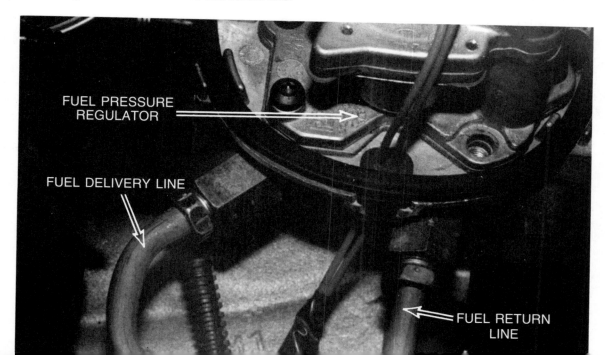

Fig. 3-4. This photograph shows the FPR, fuel delivery line, and fuel return line the way you'll find them on a GM throttle body. Remember that the fuel delivery line always has a larger diameter than the fuel return line.

body through a filter in a housing that looks like the air filter housing that sits on top of a carburetor. The volume of air entering the engine depends on the degree to which the throttle valve is open.

The throttle valve is controlled by the accelerator pedal. The farther the gas pedal is pressed, the wider the opening. The wider the opening, the greater the air flow into the engine.

What happens when the engine is idling or decelerating and the throttle valve is closed? Won't the engine flood and stall, because it's getting too much gas and not enough air? Yes—if it weren't for devices called the idle air control (IAC) and idle speed control (ISC). Which unit is used depends on the TBI system.

The IAC is screwed into an air bypass passage beneath the throttle valve. Controlled by the ECM, it has a tapered end, or pintle, to vary the size of the air bypass (Fig. 3-5). Consequently, the volume of air that's allowed into the engine through the bypass is sufficient to offset flooding at times when the driver's foot is off the gas pedal.

An ISC is a small direct-current (DC) motor that is mounted on the throttle body (Fig. 3-6). It works differently from the IAC, but with the same result. When there is no pressure on the gas pedal, the ECM activates a switch on the ISC that allows a gear to engage the throttle lever. This opens the throttle valve to allow a sufficient volume of air for the engine to run and not flood.

THE ROLE OF NON-EFI PARTS

Parts that are not members of the GM TBI system but that have a direct bearing on the performance of that system are the O_2 sensor, MAP sensor, and coolant sensor. The following descriptions clarify what these parts do and how they fit into the overall picture:

O_2 Sensor

The O_2 sensor is mounted in the exhaust system so it can keep tabs on the oxygen content of the exhaust gas. This data is supplied to the ECM, which can then "command" the injection system to alter the fuel mixture accordingly—that is, increase richness if the O_2 sensor picks up a lean condition and decrease richness if the sensor detects a rich condition. If one of the problems occur that may be caused by an overly rich or lean mixture, remember this sensor.

MAP Sensor

This sensor measures changes in pressure in the engine intake manifold that take place as the speed of the vehicle is altered and as more or less load is put on the engine. Data supplied to the ECM in the form of electrical resistance permits the ECM to vary TBI performance to meet these changes. If a problem occurs when you change speed or accelerate, remember this sensor.

Coolant Sensor

This sensor monitors engine temperature as depicted by the temperature of the coolant. Armed with this data, the ECM enables the TBI system to provide a richer fuel mixture for cold engine operation. If your car presents you with a problem only when the engine is cold, remember this sensor.

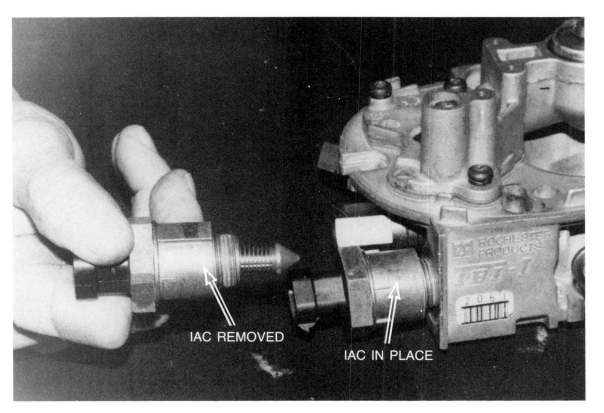

Fig. 3-5. The IAC, which is screwed into the throttle body, has a tapered end to vary the size of the air bypass and let air enter the system when the throttle valve is closed.

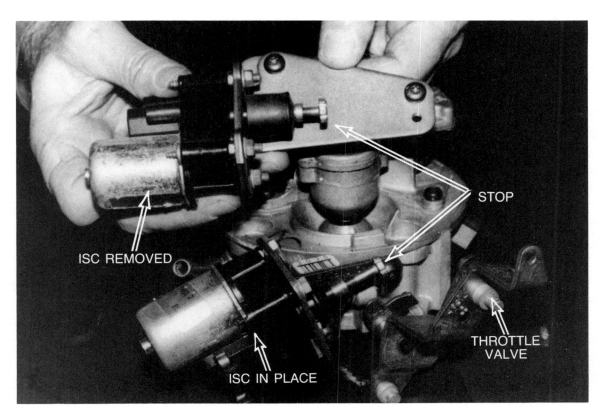

Fig. 3-6. If your GM TBI system doesn't have an IAC control, it has an ISC, which opens the throttle valve when the engine is idling.

4

GM Throttle Body Fuel Injection Systems

General Motors

Troubleshooting and Repairing Fuel Delivery Components

When the engine won't start because there might be a malfunction in the TBI system, first determine whether there is an electrical failure. *Remember:* This is an electrically controlled unit.

You do not need elaborate testing equipment to find out if electricity is present at an injector. A little instrument that you can hold in the palm of your hand and that costs less than $10.00 is all that's required (Figs. 4-1, 4-2). Called the EFI-LITE, you can buy it from Borroughs Tool and Equipment Corporation, 2429 North Burdick Street, Kalamazoo, MI 49007. Kent-Moore Tool Group, 29784 Little Mack, Roseville, MI 48066-2298 is another source.

To use the EFI-LITE, disconnect the wire harness that connects to the fuel injector electrical terminals (Figs. 4-3, 4-4). Plug the EFI-LITE into the wire harness connector (Fig. 4-5). Crank the engine. The EFI-LITE should emit pulsating flashes. If it does not flash at all or if the glow is steady and not pulsating, there is an electrical failure. Conversely, if the EFI-LITE flashes as it should, electricity is available and the engine no-start problem you're experiencing could be with the fuel delivery system. Therefore, proceed with troubleshooting.

Fig. 4-1. The easy-to-use EFI-LITE will tell you at a glance whether current is reaching fuel injectors and is pulsing the injectors to open and close them so an engine gets the amount of gas it needs to run properly.

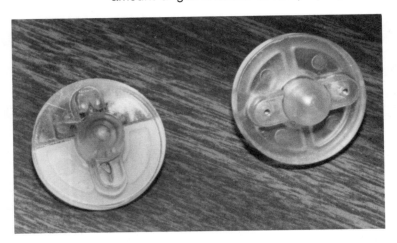

Fig. 4-2. EFI-LITE is made to fit differently configured wire connectors. When ordering the tool for your car, inform the supplier whether the injectors have pin or blade terminals, and relate the make and model of the vehicle.

Important: If there are two fuel injectors, be sure to test for the presence of electricity at both.

Fig. 4-3. To use the EFI-LITE, disconnect the wire connector from the injector. This is done by pressing the tabs.

Fig. 4-4. Pull the wire connector off the terminals of the fuel injector.

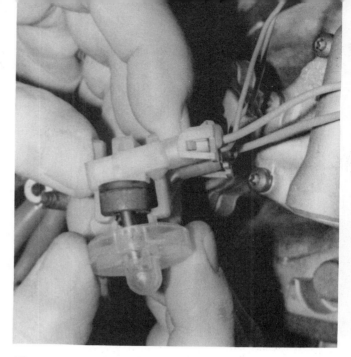

Fig. 4-5. Plug the EFI-LITE into the wire connector and crank the engine. The instrument should emit pulsating flashes. If it doesn't, there's a disruption of current.

Problems plaguing engines with TBI systems in addition to not starting include stalling, rough idling, hesitation, surging, and lack of power.

A quick test can be made of fuel delivery system performance by removing the air cleaner and having someone in the car turn the ignition key as you watch the injector or injectors (Fig. 4-6). You should see a nice steady cone-shaped stream of gas from each. No gas or what looks like an impeded supply is reason to proceed with troubleshooting.

Before you do, you should be familiar with the procedure involved in relieving fuel pressure. If you disconnect parts of the system without relieving pressure you may get sprayed with gasoline as you loosen a fitting. This can be hazardous.

To relieve fuel system pressure, follow these steps:

1. Place an automatic transmission in Park, a manual transmission in Neutral. Set the parking brake.

2. Using the information in your car owner's manual to find the fuel pump fuse, remove that fuse from the fuse panel. "Fuel Pump"

Fig. 4-6. If you observe a cone-shaped spray of gas from the tip of the injector(s), the gas delivery components of the TBI system are performing properly. Therefore, the reason for an engine performance problem lies elsewhere.

may be marked right on the panel to identify the fuse.

3. Start the engine. Even though the fuel pump is disconnected, the engine will run for several seconds before gas in the system is used up and the engine stalls. Crank the starter for another five seconds to release any residual pressure.

You can now disconnect a fuel fitting; however, be aware that even though fuel pressure has been released, a small amount of gas can still dribble out. It's a good idea, therefore, to wrap a cloth around your hand and wrench before loosening the fitting.

Before you can start the engine again, you have to reinsert the fuel pump fuse. Turn on the ignition key for several seconds to build up pressure, but do not crank the engine. Check the fitting you loosened for a leak. If gas is leaking, tighten the fitting more before starting the engine.

Important: Whenever you disconnect a fuel fitting, replace the O ring seal or seals with new ones to prevent a fuel leak.

Fig. 4-7. This illustration points out the components of the GM TBI system. Also noted is the fact that the gauge used for making the fuel system pressure test is inserted between the fuel inlet of the throttle body and the line that delivers gas to the throttle body.

THE FUEL SYSTEM PRESSURE TEST

A fuel system pressure test is the first test to make to determine if a malfunction exists within the fuel delivery system. In order to do this, you need a fuel system pressure gauge. A gauge designed for testing the fuel pressure of GM TBI systems is sold by both Kent-Moore and Borroughs.

GM TBI systems develop pressures between 9 and 13 psi. If you don't have a service manual for your car that contains the exact specification, but find that the pressure when you do the test is between 9 and 13 psi, you can dismiss parts of the fuel delivery system as being the cause of your problem. These parts are the fuel pump, fuel pump relay, fuel sock, fuel line, and fuel filter (Fig. 4-7). The trouble will be found instead with a part of the throttle body, including the injector, or with another system of the vehicle.

To test fuel pressure, follow these steps:

1. Relieve fuel system pressure (see above). Remove the air cleaner assembly from the throttle body. Notice the vacuum port on the throttle body to which the vacuum hose from the air cleaner connects. Insert a golf tee or stub of a pencil into the end of this port to prevent a vacuum loss when cranking the engine as you test fuel system pressure. A vacuum loss may cause a false test result.

2. Remove the gasket that lies under the air cleaner (Fig. 4-8). Bring this gasket to the

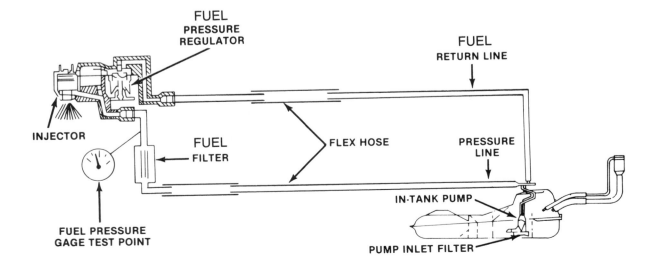

parts department of a GM dealer to ensure getting a replacement that matches. After you have the correct gasket, discard the old one.

3. Loosen the fuel line, which is screwed into the fuel inlet, and pull it free (Fig. 4-9). There are *two* lines connected to the fuel meter assembly. The one having the larger diameter is the fuel inlet; the other one is the fuel return. The fuel inlet is the one that brings gas to the fuel injector. The fuel return diverts excess gas away from the injector to the gas tank. If you aren't sure which is which, trace the line. The one that has a fuel filter spliced into it is the fuel inlet.

4. Connect the fuel pressure gauge between the fuel inlet line and the fuel inlet fitting of the throttle body (Fig. 4-10).

Fig. 4-8. GM recommends that whenever you remove the air cleaner assembly from the throttle body you replace the gasket that seals the point where the two meet.

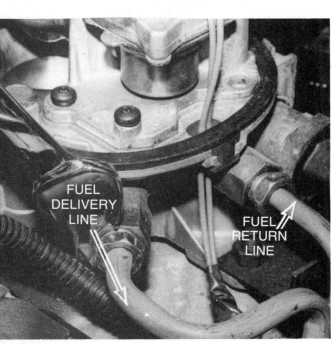

Fig. 4-9. Unscrew the fuel delivery line at the throttle body. The fuel delivery line has a larger diameter than the fuel return line. If you can't distinguish between the two, trace the lines. The in-line fuel filter is located in the fuel delivery line.

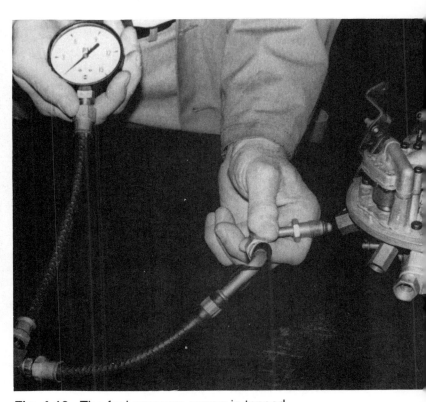

Fig. 4-10. The fuel pressure gauge is tapped into the system between the fuel delivery line and fuel inlet fitting of the throttle body.

5. Make sure the fuel pump fuse you removed when relieving fuel pressure has been reinstalled; then turn on the ignition switch, but don't crank the engine. Observe the reading on the gauge within two seconds after turning on the switch (Fig. 4-11). Then turn off the switch. You will get one of four results: (a) a normal reading; (b) a zero reading; (c) a lower than normal reading (less than 9 psi); (d) a higher than normal reading (above 13 psi). If you get a normal reading, indicating that the components involved in delivering fuel to the throttle body are in good shape, there are the parts of the throttle body to consider (Chapter 5).

6. When you've completed the fuel system pressure test, relieve fuel system pressure again before disconnecting the fuel pressure gauge. Then, reconnect the fuel inlet line, turn on the ignition, and check for leaks.

ZERO FUEL PRESSURE READING

If you get a zero fuel pressure reading, the engine won't start. Malfunctions that cause this are a clogged fuel filter, a bad fuel pump, or a breakdown in the fuel pump electric circuit.

The first thing to do is inspect the fuse that protects the fuel pump circuit. Replace it if you see that the element is broken. If the engine now starts but the fuse blows again, there is a short in the fuel pump circuit. Consult an auto electrical specialist if you don't know how to find and fix shorts.

A zero pressure reading can be caused by a clogged sock or in-line fuel filter. The sock is also a filter (Fig. 4-12). It is inside the gas tank and is responsible for filtering large particles and water from gas.

Fig. 4-11. The reading seen on the fuel pressure gauge will allow you to assess whether there is a malfunction in the system that delivers gas to the throttle body.

Fig. 4-12. When an engine that won't start is accompanied by a zero fuel pressure reading, the electric fuel pump and fuel sock become suspects. The fuel pump and sock are in the fuel tank. If the sock is clogged, but the pump is good, the sock can be removed from the pump for cleaning or replacing.

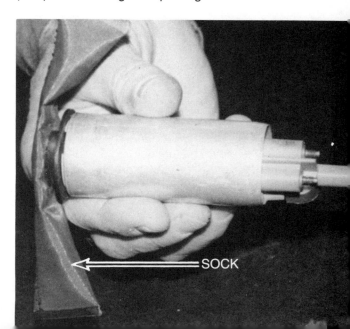

Fig. 4-13. There are several relays in the engine compartments of GM vehicles. This group of three is typical of what is found. Use the electrical diagram in the shop manual to pinpoint the fuel pump relay. Another way to find the relay is to disconnect one relay at a time and listen at the fuel tank as someone in the car turns on the ignition key. When you don't hear the fuel pump whir, you've found the fuel pump relay. Disconnecting the relay puts the fuel pump out of action.

To get an idea if the sock or in-line fuel filter is clogged, turn off the ignition switch for a few minutes, put your ear near the fuel tank, and have someone turn the ignition switch back on. Do not crank the engine.

Do you hear a whirring? If so, the fuel pump is working, which means the sock or in-line filter may be clogged. See below for a discussion of how to handle filters.

If the fuel pump isn't whirring, check the fuel pump relay before deciding that the fuel pump is bad and has to be replaced. In most GM vehicles having TBI, the fuel pump relay is in the engine compartment (Fig. 4-13). Get rid of corrosion by disconnecting wires and cleaning terminals with a piece of emery cloth; then reconnect wires securely.

Does the engine start? Still not? Then, replace the relay. It may not be bad, mind you. But on the chance that it is, the relay is a lot less expensive to replace than a fuel pump. If a new relay fails to produce that whirring sound you want to hear, it's time to replace the fuel pump. The fuel tank has to be removed from the car and opened to do this. If you aren't comfortable with this procedure, let a mechanic do the job for you.

LOW FUEL PRESSURE READING

A clogged in-line fuel filter will give a low fuel pressure reading more often than a zero reading. A filter that is plugged restricts the flow of gas to the injectors, resulting in a reduction in fuel pressure. The engine will not start or will start and die.

Depending on the GM model car you're working on, the in-line filter is located under the rear of the vehicle attached to the frame, against the outside of the fuel tank, or in the

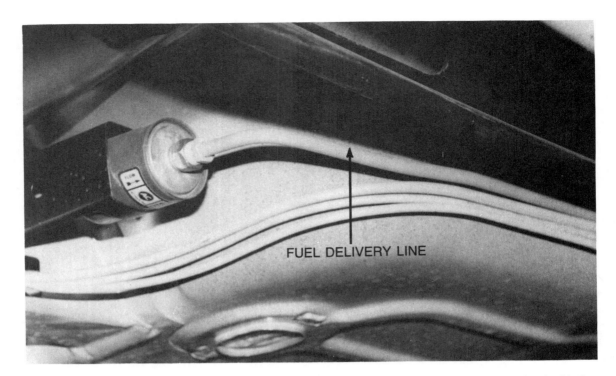

Fig. 4-14. The in-line fuel filter of GM TBI systems filters gas as it flows to the throttle body. Notice the three fuel lines running alongside the pump. One is the fuel return line; the other two are connected to the charcoal canister of the fuel evaporation emissions control system. When an engine problem arises, make sure all lines are in good condition. Replace one if it is cracked or crushed.

Fig. 4-15. To replace the in-line fuel filter, relieve pressure from the fuel system before unscrewing the filter.

Fig. 4-16. When you put the new filter in place, make sure the flow arrow points toward the throttle body—that is, in the direction in which gas flows.

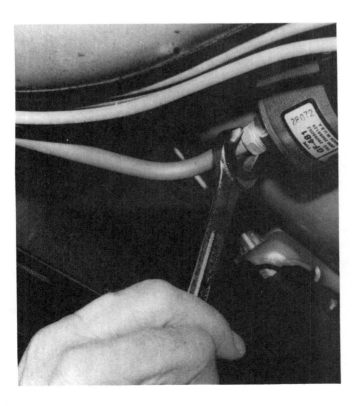

front of the engine compartment adjacent to one of the tire and wheel assemblies. If you can't spot the filter easily, trace the fuel inlet line back from the throttle body (Fig. 4-14). Sooner or later, you'll run into the filter. To replace it, relieve fuel pressure and unscrew the fuel line from each side of the filter (Fig. 4-15).

When you buy a new filter, be sure seals come with it. Install the seals between the filter and fuel line fittings. By hand, screw the fittings to the filter, with the flow arrow pointing toward the throttle body (Fig. 4-16). Then, tighten the fittings (Fig. 4-17), start the engine, and check for leaks. If there is a leak, tighten the fitting a bit more.

Another thing that can cause low fuel pressure is a restricted fuel inlet line, but it's not likely for dirt to be the reason since the fuel line is protected from dirt by the sock and in-line fuel filter.

The fuel inlet line is more apt to incur damage, so inspect the line from the fuel tank to the throttle body. If you find that it's kinked, replace it.

If you still haven't found the reason for low fuel pressure, remove the fuel tank and take out the in-tank fuel delivery assembly. The sock is probably plugged. Clean or replace it.

HIGH FUEL PRESSURE READING

Suppose you get a fuel pressure reading in excess of 13 psi. It doesn't happen often, but when it does the reason for it is clear. Look to the fuel return line. That line is kinked somewhere between the throttle body and fuel tank. If gas can't flow back from the throttle body to the fuel tank, high pressure is created within the system.

That's it as far as components that deliver gas to the throttle body are concerned. But as mentioned before, trouble with a GM car's

Fig. 4-17. Tighten the fittings. Then start the engine and let it run for a second or two before turning it off. Now run your finger around each fitting to make sure there is no leak.

performance can be caused by damaged parts attached to the throttle body. What to do about this is the subject of the next chapter.

5

GM Throttle Body Fuel Injection Systems

General Motors

Servicing the Throttle Body

Fig. 5-1. This photograph will help you identify the various parts of a GM throttle body that are discussed in this chapter.

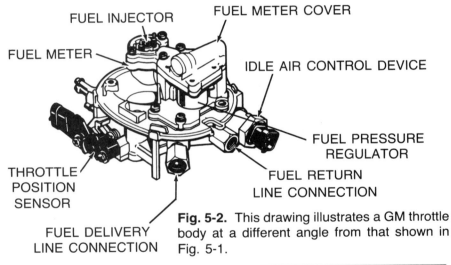

Fig. 5-2. This drawing illustrates a GM throttle body at a different angle from that shown in Fig. 5-1.

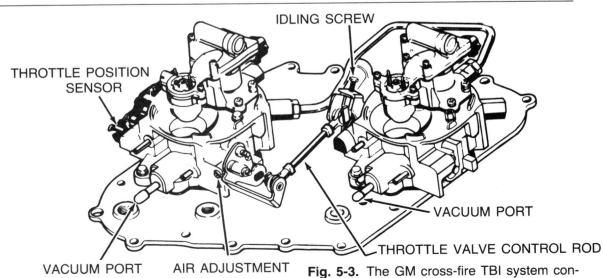

Fig. 5-3. The GM cross-fire TBI system consists of two single-point TBI units coupled together to work in tandem.

LOOKING AT FUEL INJECTORS

When a TBI system part malfunctions, the performance problem it creates is generally associated with that part. For example, a damaged fuel pump relay is responsible for hard starting. This makes troubleshooting quite simple.

Let's begin the discussion with fuel injectors. If an injector is stuck in the partly open position, dieseling may occur. Dieseling refers to an engine that keeps running after the ignition is turned off. When you turn off the ignition, some fuel still flows into the cylinders through the partly open injector. If the engine is hot enough, fuel is ignited and the engine chugs on until the excess is consumed.

Whether installed in a single-point, dual-point, or cross-fire GM TBI unit, fuel injectors are removed, inspected, and replaced the same way (Figs. 5-1, 5-2, 5-3). If you replace an injector, make sure you get the correct one for your system. Injectors are calibrated differently from one system to another.

To remove a fuel injector from a GM throttle body, do the following:

1. Relieve pressure in the fuel system.

2. Remove the air cleaner and discard the air cleaner gasket. Get a new one.

3. Detach the electric connector from the fuel injector terminals.

4. Unscrew the fuel meter assembly cover (Fig. 5-4). Notice if the injector is held by a metal bracket. If so, remove the screw in order to free the bracket.

5. Get something to use as a fulcrum—the shank of a small screwdriver or a round piece of metal stock, such as a welding rod, will serve. Also, get a screwdriver with a fairly long handle. Place the fulcrum across the fuel meter body assembly, put the long-handled screwdriver over it so the tip of the screwdriver engages the ridge that's around the fuel injector, and press down to pop the injector out of the throttle body (Figs. 5-5, 5-6).

Fig. 5-4. To get at the fuel injector, remove the fuel meter assembly cover.

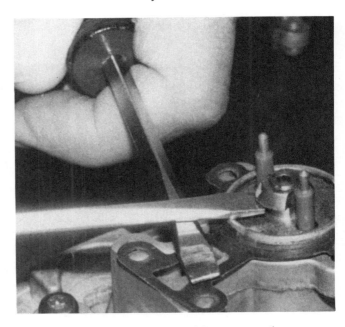

Fig. 5-5. Using a fulcrum and lever, pop the fuel injector from its seat.

Fig. 5-6. When the injector is loose, lift it out of the throttle body.

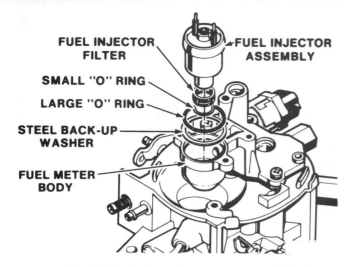

Fig. 5-7. At first glance, you may think there are a number of fuel injector seals that have to be replaced, but once a fuel injector is taken from the throttle body only the small and large O rings should be discarded for new ones when the old injector is reinstalled or a new injector is used.

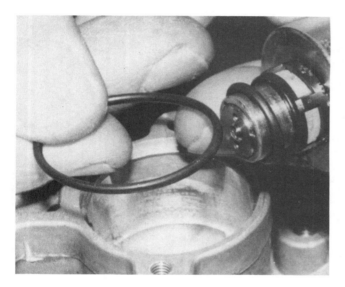

Fig. 5-8. Here is what fuel injector O rings look like. The large O ring is being held. The small O ring is still in place on the end of the injector.

Fig. 5-9. Inspect the nozzle end of the injector. If you see rust, discard the injector.

Note: The injectors of the GM model 700 TBI system can be removed by grasping them with your fingers and pulling. If you don't know which model unit the car possesses, try the hand method before resorting to the fulcrum–screwdriver procedure.

6. When the injector is out of the throttle body, retrieve the O rings that act as seals (Figs. 5-7, 5-8). One O ring may be around the upper part of the injector; the other may be around the lower part. Discard both.

You might be able to tell almost immediately if the fuel injector has to be replaced by examining the nozzle (Fig. 5-9). Do you see dirt or what looks like rust? If you do, get a new fuel injector. It's a waste of time immersing the part in a liquid cleaner in the hope of salvaging it, because the liquid cleaner will damage the electric parts of an injector.

If dirt or rust is visible on the injector, you have to deal with the source of the contamination. It is probably coming from the fuel tank, which should be removed and cleaned by a shop that specializes in the service.

Suppose there is no dirt or rust visible on the tip. You now have a choice to make. If you have a fuel injector tester to determine fuel flow and spray pattern, use it. Otherwise, seek professional assistance or just replace the injector.

When you reinstall the old injector or put in a new one, follow this procedure:

1. Lubricate new upper and lower O rings with automatic transmission fluid, such as DEXRON II. Install O rings. Apply lubrication to the lower end of the injector to facilitate installation (Fig. 5-10).

2. With all injectors except those from model 700 units, align the pins on the injector with the hole in the fuel meter body (Figs. 5-11, 5-12, 5-13) and push the injector straight

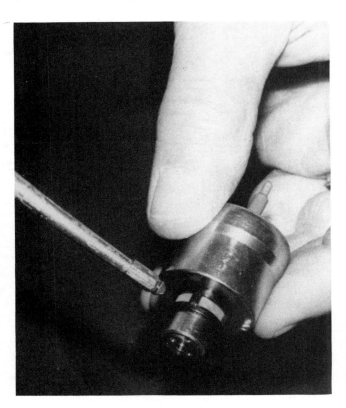

Fig. 5-10. Lubricate the end of the injector to make it easier to install the part. Use automatic transmission fluid, such as DEXRON II.

Fig. 5-11. Examine the injector for any alignment pins. You may find one on the side of the part.

Fig. 5-12. You may also find an alignment pin on the bottom of the injector housing.

Fig. 5-13. Be sure to line up the pins on the injector so they intercept their respective seats in the injector housing. Doing so insures that the injector is correctly placed.

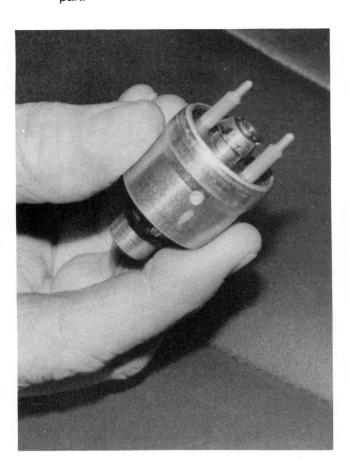

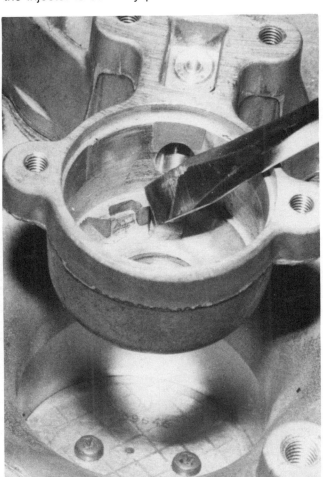

into the cavity (Fig. 5-14). Injectors of model 700 units don't have pins. Just push them back into place.

3. Reattach the bracket. It's recommended that you smear the threads of the bracket screw with a thread-locking compound, such as Loctite 262. This compound should be applied to all screws you remove from the throttle body, except the threaded parts of the IAC and throttle position sensor.

4. Reinstall the fuel meter assembly cover and reattach the fuel line and electric connector (Fig. 5-15).

5. Turn the ignition switch on, but do not crank the engine. Wait a few seconds and check for fuel leaks.

WHEN THE IDLE AIR CONTROL OR IDLE SPEED CONTROL IS SUSPECT

If the IAC or ISC fails, the symptoms are an engine that exhibits a rough, unstable idle. If the condition is bad enough, the vehicle may stall. A faulty IAC or ISC will display a code 35 on the CHECK ENGINE light of the vehicle's instrument panel. If you don't know how to get the light to reveal the code, have a mechanic do it. Other than this, the only way to test an IAC or ISC is with a substitute to see if it puts an end to rough idling.

To replace an ISC, disconnect the wire and unbolt the motor and its bracket from the throttle body (Fig. 5-16). To replace an IAC, disconnect the wire and unscrew the device from the throttle body (Fig. 5-17).

Fig. 5-14. Press the injector into the injector housing to seat the part securely.

Fig. 5-15. Before you reinstall the throttle body, move the throttle lever back and forth by hand to make sure that the throttle valve works freely.

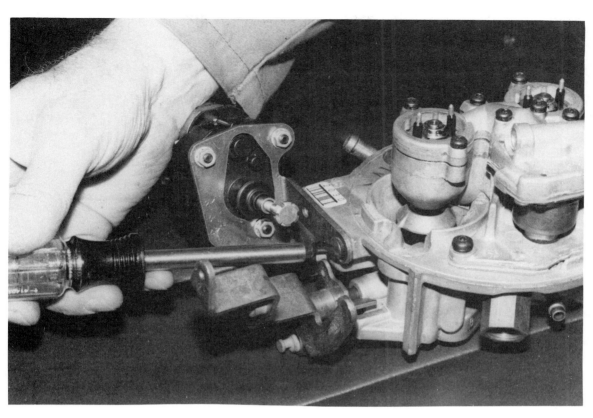

Fig. 5-16. If the ISC motor becomes suspect, it is easily replaced.

Fig. 5-17. The IAC is also easily removed. Loosen it using a wrench; then unscrew it from the throttle body.

Note: GM uses several different kinds of IACs, so make certain the one you get as a replacement is the one needed by your particular TBI unit.

THE THROTTLE POSITION SENSOR

There is one part of a GM TBI assembly not yet described: the throttle position sensor (TPS).

The TPS is an electrical device called a potentiometer. It translates the angle of the throttle valve opening into an electrical signal, which is transmitted to the ECM. The purpose is to help the ECM calculate the ratio of air to gas needed by the engine.

Suppose the TPS goes bad: What happens to engine performance? It often goes to pot, because the air-to-gas ratio is seriously disrupted. Every condition that affects an engine when it's forced to operate on an overly rich fuel mixture can occur. This includes rough engine idle, stalling, hesitation, black exhaust smoke, and poor fuel economy. Replace the TPS when a condition occurs that may have an overly rich fuel mixture as its cause.

The TPS of your GM TBI system is probably held to the side of the throttle body with screws, although some units have been welded into place. If the latter, you have to drill the welds out, tap the holes, and replace the

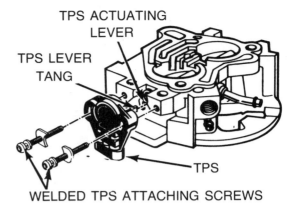

Fig. 5-18. Replacing a TPS that is welded in place can be a nuisance. Welds have to be drilled and tapped to break the welds. A TPS that uses screws can then be used as a replacement.

part with one that screws into place (Fig. 5-18). This may not be a job you want to tackle.

Fortunately, except for early models, most GM TBI units use screw-on TPS, which are easy to replace. Remove the air cleaner and air cleaner gasket. Replace the gasket with a new one. Disconnect the connector and unscrew the TPS from the throttle body (Fig. 5-19). Install a new TPS of the same type.

Note: You do *not* have to relieve fuel pressure to replace the TPS, but to be safe disconnect the battery ground cable.

Fig. 5-19. Fortunately, the TPS on most GM vehicles are just screwed on using nonwelded fasteners. Use a Torx driver to remove screws.

THE FUEL PRESSURE REGULATOR

The FPR is set at the factory. Seldom does anything go wrong with it, but when it does the reason is usually a tear in the diaphragm that helps to maintain constant pressure. A damaged FPR may leak gasoline.

Fuel pressure is regulated by the difference between fuel pump pressure acting on one side of the diaphragm and the force of a calibrated spring acting on the other side. When fuel pressure drops below what it should be, the spring forces the diaphragm up toward an opening to block the return of fuel to the fuel tank through the fuel return line. If the diaphragm fails, the engine may get too much gas and flood.

The FPR in a typical GM TBI system is an integral part of the fuel meter body (Figs. 5-20, 5-21). To replace the part, it is necessary to get a new fuel meter cover (Fig. 5-22). The FPR of the model 700 TBI, however, can be repaired by installing the diaphragm sold by GM dealers in a repair kit.

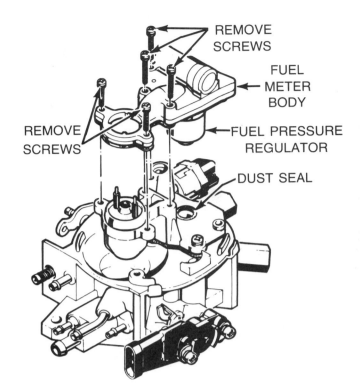

Fig. 5-20. The FPR in many GM vehicles is an integral part of the fuel meter assembly. If the FPR goes bad, the entire assembly may have to be replaced.

Fig. 5-21. To get at the FPR, unscrew the fuel meter. Notice the dust seal in the hole on which the FPR sits. Be sure the seal is in place when reinstalling the unit.

Fig. 5-22. If you aren't sure whether the entire fuel meter has to be replaced or if there's a rebuild kit available for the FPR, ask an auto parts dealer.

6

GM Multiport Fuel Injection System

General Motors

Preliminary Troubleshooting and Repair

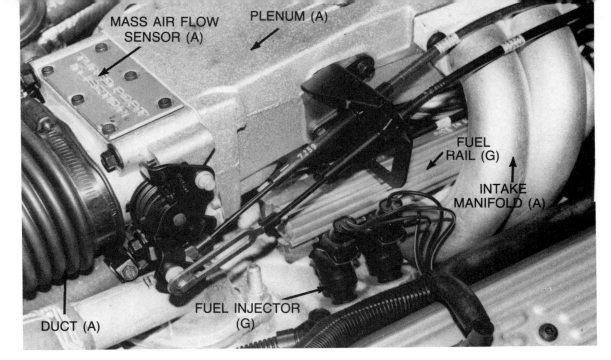

Fig. 6-1. Some key parts of the gasoline and air circuits of a typical GM MPFI system are identified in this illustration. They are labeled "G" for gas and "A" for air.

A MULTIPORT FUEL injection (MPFI) system is also referred to as a multipoint fuel injection system, and port fuel injection (PFI) system. Whatever it's called, the system consists of two circuits—a gasoline circuit and an air circuit—and one fuel injector for each cylinder the engine possesses (Fig. 6-1). Thus, unlike GM's throttle body injection (TBI) systems (Chapters 2 to 5), which have at most two injectors, a GM port injection system has eight injectors serving an eight-cylinder engine, six injectors for a six-cylinder engine, and four injectors for a four-cylinder engine.

The fuel injectors are part of the gasoline circuit (Fig. 6-2). The route taken by gas from the fuel tank is through a filter (called a sock) that's in the gas tank, into and out of the fuel pump, to another filter (called an in-line filter), which is spliced into the fuel line, to the fuel pressure regulator (FPR), and to the injector.

The injectors spray gas into the engine's fuel intake system at the entry side of the intake valves (Fig. 6-3). It's at this point that gas mixes with air, which enters the intake

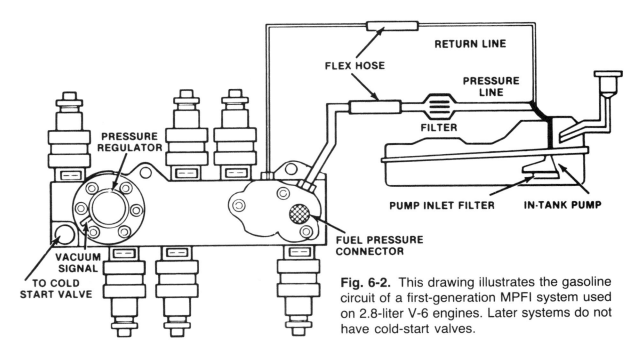

Fig. 6-2. This drawing illustrates the gasoline circuit of a first-generation MPFI system used on 2.8-liter V-6 engines. Later systems do not have cold-start valves.

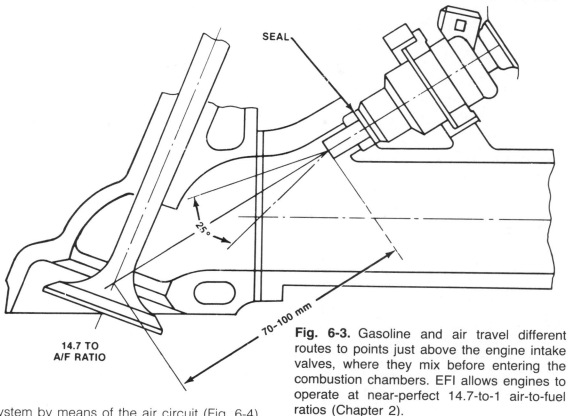

Fig. 6-3. Gasoline and air travel different routes to points just above the engine intake valves, where they mix before entering the combustion chambers. EFI allows engines to operate at near-perfect 14.7-to-1 air-to-fuel ratios (Chapter 2).

system by means of the air circuit (Fig. 6-4). The mixture then passes through the intake valves into the cylinders and is burned to power the vehicle. It's as simple as this.

As with an engine equipped with a carburetor, a malfunction in any component of the gas or air circuit of a GM MPFI system disrupts the fuel mixture and sets engine performance on its tail. When there is trouble in a circuit, the engine demonstrates one or more of the following problems: It starts and stalls, won't start at all, lacks power, idles poorly, hesitates, surges, and stumbles.

Fig. 6-4. The air circuit of a GM MPFI system is illustrated. Air flows into the engine through the air cleaner, air duct, mass air flow sensor (which incorporates the throttle valve), air plenum, and intake manifolds.

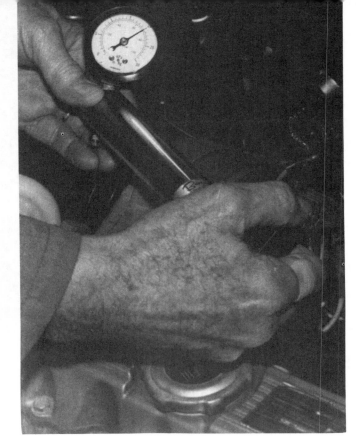

Fig. 6-5. Don't lose sight of basics: Remember that engine performance problems are often caused by vacuum loss; so test vacuum before delving into an MPFI system.

Fig. 6-6. Inexpensive and easy to use, the EFI-LITE tells at a glance if there is a disruption of electricity to a fuel injector.

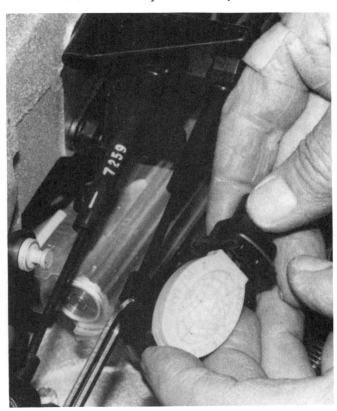

When one of these conditions arises, you have to troubleshoot the system by checking the gas and air circuits separately until you find the misadjusted or bad part causing the problem. In most cases, however, when a GM engine develops a problem that is the fault of the MPFI system, the cause of the trouble lies in one of two places: with an electrical failure or with the fuel injectors. It is with these that you should begin after making sure that there is no disruption in vacuum or a problem with an electrical connector (Chapter 1) (Fig. 6-5).

MAJOR CONDITION NO. 1: NO JUICE

The fuel injectors of a GM MPFI system are electrically operated solenoid valves. If current is not present at the injectors because of an electrical failure, the injectors won't open; therefore, the engine won't get gas and won't start. Consequently, if your engine doesn't start, determine if the problem is being caused by an electrical failure.

You don't need a high-powered, expensive analyzer to find out if there's a disruption of electricity to the fuel injectors—just an instrument you can hold in the palm of your hand called an EFI-LITE (see Chapter 4, page 24). To use the EFI-LITE, pull apart the electrical harness connector from any injector. There are two parts to a connector—one end attaches to a branch wire from the main harness, and the other plugs into the injector.

Plug the EFI-LITE into that part of the connector which is attached to the branch wire—not into the one on the injector side (Fig. 6-6). Crank the engine. If the EFI-LITE flashes, it means electricity is getting to the injectors. The malfunction that is keeping the engine from starting is therefore not located in the car's electrical system.

On the other hand, if the EFI-LITE doesn't glow or if it emits a steady rather than a pulsating beam as the engine is cranked, there's probably trouble in the main circuit. But don't take one injector's "word" for it. The branch wire feeding current to this particular injector from the main harness may be damaged, so trusting one test can condemn the entire electrical system for the fault of a short length of wire. Therefore, test for electricity at another injector. If you get the same result, then an electrical failure is the reason your engine isn't starting.

Note: If you have little knowledge of the electrical system, leave troubleshooting and repair to a professional technician.

MAJOR CONDITION NO. 2: DIRTY FUEL INJECTORS

Among the persistent performance problems caused by dirty fuel injectors in a GM MPFI system are hard starting, rough idling, hesitation or stumbling while accelerating, surging, stalling, and a lack of power. Dirt inside one or more fuel injectors reduces the amount of gas reaching the engine, thereby resulting in a lean fuel mixture. But don't be too fast to replace the injectors. You may be able to get rid of the problem much less expensively by cleaning the injectors.

Try switching to a premium unleaded gas. If two or three tankfuls fail to clear the condition, pour a can or two of fuel injector cleaner into the fuel tank.

Assuming these easy-to-apply procedures don't work, the next course of action is a fuel injector balance test, which will confirm whether injectors are really dirty (or damaged) or if the drivability problem is being caused by something other than the injectors.

To do a fuel injector balance test, you need a GM Multiport EFI System Diagnostic Tester, which sells for about $200. This is made by Kent-Moore and carries tool number J-34730-A (see page 24 for address).

The GM Multiport EFI System Diagnostic Tester includes a fuel pressure gauge and a remote control unit that allows you to operate the injector. Naturally, if you don't want to bear the cost of the tester, a professional mechanic can do the test for you. A fuel injector balance test on a GM engine having an MPFI system is the only way other than indiscriminately replacing injectors to establish if injectors are dirty (or damaged).

The test is done as follows:

1. Locate the fitting for the fuel pressure gauge on the fuel rail, and connect the tester's gauge to it. The fuel rail is the elliptical metal pipe around the engine to which all injectors connect. It's through this rail (fuel line) that gas is delivered to the injectors. The fuel rail fitting resembles a tire valve. It should have a cap over it to protect it from dirt (Fig. 6-7). Unscrew the cap (Fig. 6-8).

Fig. 6-7. Every GM engine with MPFI has a fuel pressure gauge fitting (arrow) on the fuel rail. It's at this point that testing is done. Be sure to keep the cap over the fitting when you aren't doing a test to prevent dirt from getting into the fuel rail.

Fig. 6-8. When the cap over the pressure gauge fitting (arrow) is removed, the fitting is ready to accept the fuel pressure gauge.

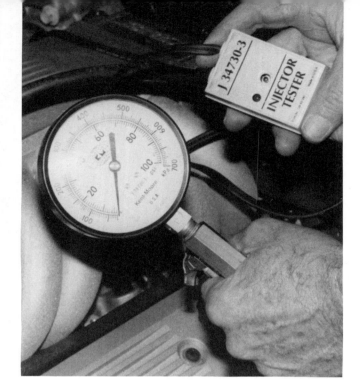

Fig. 6-9. The two instruments that are needed to test fuel injectors of a GM MPFI system are a fuel pressure gauge and a remote control unit that activates the injector being tested.

Fig. 6-10. This instrument can be used to force-clean the fuel injectors in any model car—not only GM—as long as there is a fuel pressure gauge fitting available to connect the instrument. Since dirty injectors are more likely to occur than damaged injectors, trying the cleaning procedure is worth the effort before you remove the fuel injectors of an MPFI system, which could be a long, involved, and expensive process.

Caution: Wrap a towel around your hand and the fitting to catch gas that may spray as you attach the gauge. Obviously, don't smoke or have anything else around that can cause a spark.

2. After the gauge is attached, pull apart the connector at an injector and attach the remote control unit to the part of the connector on the injector. Connect the cables from the remote control unit to the positive and negative terminals of the car battery. The power needed for the test is provided by the battery.

3. Wrap a towel around the vent valve of the fuel pressure gauge to catch any gas that might leak. Start the engine; then open the valve to expel air trapped in the gauge. Close the valve and shut off the engine.

4. Let the engine stay off at least 10 seconds.

5. Start the engine again and press the button on the remote control unit *one* time only (Fig. 6-9). The needle on the pressure gauge should soar, settle back, and stop. Note at which numeral the needle stops.

6. Repeat the test for all the injectors, being certain to reconnect each injector to the main harness after doing so.

When you have taken all the readings, compare them. Each should be within 10 kPa (kilopascals) of the others. If one injector shows a lower reading than the others, it means that injector is either dirty or damaged. Considering that dirty fuel injectors are more likely to exist than damaged injectors, it is worth the effort to force-clean the injectors before replacing them.

Incidentally, 10 kPa is equivalent to 1.65 psi. The pressure gauge is marked off in both scales, but the metric scale (kPa) is easier to read.

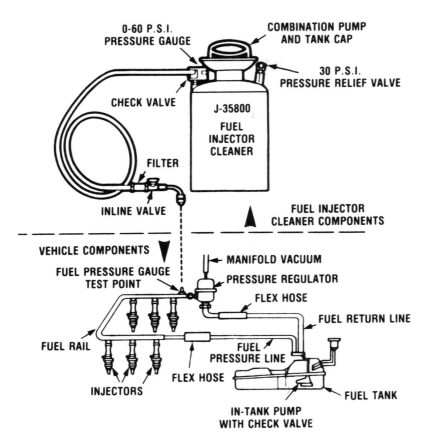

Fig. 6-11. The fuel injector cleaner sends the cleaning agent through the injectors, dislodging dirt.

FORCE-CLEANING INJECTORS

To force-clean injectors, there is a fuel injector cleaning instrument that costs about $150 (Fig. 6-10). It is effective and easy to use since you don't have to remove the injectors from the engine.

The cleaning instrument is a 2-gallon steel tank that's outfitted with fuel-resistant hoses and a gauge (Fig. 6-11). To force-clean injectors, mix and pour 1½ quarts of gas and 3 ounces of an injector cleaner into the 2-gallon tank. An injector cleaner can be ordered from Borroughs (part number BT8606) or Kent-Moore (part number J-358000) or purchased from a GM dealer (see page 24 for addresses). The GM product is called GM Top Engine Cleaner or AC/Delco X66-P.

Once the mixture is in the tank, follow this procedure:

1. Attach the hose from the tank to the fitting on the fuel rail as you did in performing the fuel injector balance test (Fig. 6-12).

2. Pump the handle of the tank until the gauge on the tank shows a reading of 25 psi.

3. Open the hose inlet valve a half turn.

4. Disable the car's electric fuel pump by pulling the fuel pump fuse or disconnecting the fuel pump relay (pages 25–26).

5. Start the engine and run it at about 2,000 rpm until the cleaning agent is consumed and the engine stalls. This takes about 15 minutes if the engine has four or six cylinders; about 10 minutes if the engine is a V-8. When the engine stalls, turn off the ignition.

6. Close the hose inlet valve and disconnect the hose from the fuel rail fitting. Replace the fuel pump fuse or reconnect the fuel pump relay.

7. Start the engine and let it run at a speed between 1,000 and 2,000 rpm for 5 minutes to let the cleaning agent flush itself from the fuel rail.

8. Test drive the car to see if the engine performance problem has been resolved. If it hasn't, replace the injector or injectors that didn't pass the fuel injector balance test.

Fig. 6-12. In this illustration the technician is getting ready to clean fuel injectors by attaching the instrument to the fuel pressure gauge fitting.

7

GM Multiport Fuel Injection System

Troubleshooting and Repairing Other Components

Fig. 7-1. Remember to disconnect the battery ground (negative) cable for safety. You can do this at the battery or where the cable attaches to the engine.

IN DEALING WITH GM MPFI, remember that fuel injectors (Chapter 6) are only part of the fuel delivery system. However, they are frequently blamed for problems that lie with other components. In troubleshooting these other components, the following precautions should be observed:

- Unless otherwise instructed, the negative cable should be kept disconnected from the car battery (Fig. 7-1).

- Keep a dry chemical (class B) fire extinguisher close by.

- When loosening or tightening fittings, use the double-wrench technique—that is, one wrench to hold one side of the fitting steady and the other wrench to do the turning.

- All fittings used on parts that deliver gas to the injectors have O ring seals to prevent leaks. When you disconnect a fitting, discard the old O ring and replace it with a new one of the same design (Fig. 7-2).

- If you have to replace a metal fuel pipe, make certain it meets GM specification 124-M. If you have to replace a rubberized fuel hose, make certain it meets GM specification 6163-M.

- Never replace a metal fuel pipe with a fuel hose.

- To reduce the risk of fire and injury, make sure you relieve pressure in the fuel system before servicing any part of the system.

Fig. 7-2. Lines connected to parts that handle gas have O ring seals. When you disconnect a line, discard the old seal and replace it with a new one.

RELIEVING FUEL SYSTEM PRESSURE

To relieve pressure in the fuel delivery system, cover the fitting on the fuel rail with a cloth; then connect a fuel pressure gauge to the fitting (Fig. 7-3).

The gauge has a bleed valve. Attach a hose to it, aim the other end of the hose into a container, and open the bleed valve to release fuel system pressure. As pressure drops, the gauge needle will also drop. When the needle hits zero, all pressure has dissipated and it's safe to work. But remember: Turning on the ignition switch will cause pressure to build up. If this is done, you must relieve pressure again if servicing is to continue.

Once pressure is released, proceed to service parts of the GM MPFI system that may be mechanically deficient and causing a performance problem. The remainder of this chapter offers a logical troubleshooting procedure (Fig. 7-4).

Fig. 7-3. The fuel pressure gauge is an important tool for working on a GM MPFI system. Use it to relieve fuel system pressure as well as to determine what is wrong with a system. Notice the bleed valve (arrow).

Fig. 7-4. Use this drawing of a typical GM MPFI system as a guide when troubleshooting. If your engine has a cold-start fuel injector and thermo-time switch, suspect that one or both have failed if the problem you're having is hard starting, rough idling, or stalling only when starting a cold engine. Most GM engines with MPFI systems do not have these cold-start parts.

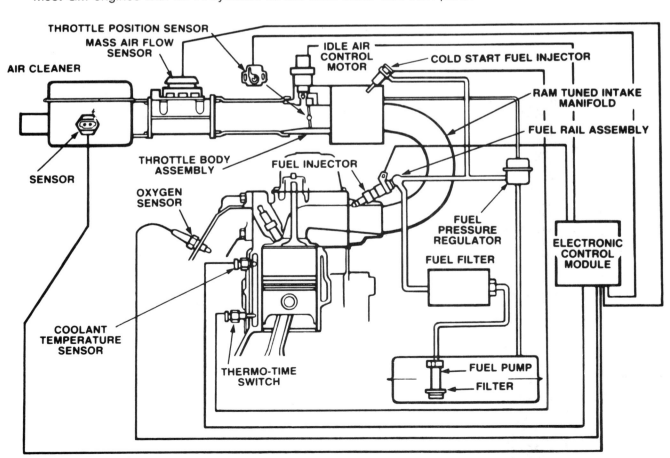

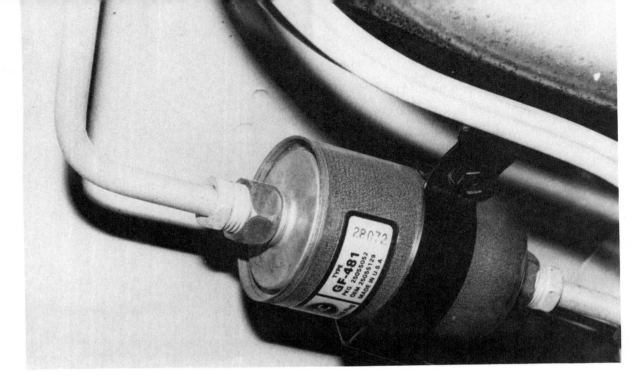

Fig. 7-5. The in-line fuel filter in most GM vehicles with multiport fuel injection is secured in a bracket at the outlet side of the fuel tank. Gas pumped into it is filtered before proceeding to the fuel injectors.

HANDLING FUEL FILTERS

If a fuel filter is clogged, the engine won't start, or it will start and die. Two kinds of fuel filters are installed in the fuel delivery systems of every GM engine that has MPFI: (1) an in-line filter that is spliced into the fuel pipe, which extends from the fuel tank to the fuel rail; (2) a fuel filter called a sock in the gas tank, which is attached to the fuel pickup tube.

The purpose of both filters is to prevent dirt from getting into the fuel delivery system. The sock also stops water from entering the system.

Large dirt particles and water should be stopped cold by the sock. Under normal circumstances, the sock is self-cleaning and requires no maintenance. If it does get plugged, however, gas can't get through; then it has to be replaced and the fuel tank cleaned.

Replacing the sock and cleaning the fuel tank are jobs for a professional garage. Therefore, consider doing other servicing first, until little doubt remains that the sock is clogged.

The responsibility of the in-line fuel filter is to act as a backup to catch smaller particles of dirt that escape the coarser mesh sock (Fig. 7-5). Replacing an in-line filter is easy and should be a high-priority job if the engine develops a starting problem.

In most GM cars sporting MPFI, the in-line filter is found on a rear crossmember, just in front of the gas tank. If it isn't there, look for it elsewhere in the line that carries gas to the fuel injectors.

After releasing fuel pressure, disconnect the fuel line fitting from each end of the filter (Fig. 7-6). Unscrew the filter from the crossmember, throw it away, and install a new filter (Fig. 7-7). Be sure to use new O rings to seal fuel line-to-filter fittings.

UNDERSTANDING THE FUEL PUMP

The fuel pump used by GM cars with MPFI systems is electrically operated and is located inside the fuel tank. If the fuel pump fails, gas cannot get to the engine and the engine won't start.

The quickest way to determine if the fuel pump is working is to do a "hearing" test. Turn off the ignition switch and go to the rear of the car. Have someone in the car turn on the ignition switch. If you don't hear whirring from the fuel tank, the pump has either seen better days, or there's a breakdown in the circuit that energizes the pump.

Begin with the circuit, which consists of a relay, wiring, and a fuse. On a chance that the fuse protecting the circuit has blown, replace it. Check the car owner's manual to identify where in the fuse panel the fuse is located. If the engine now starts but dies again because the new fuse blows, there's a short in the circuit that a GM electrical specialist should be called on to fix.

Next, turn your attention to the relay. When you first turn on the ignition switch, the electronic control module (ECM) turns on the relay for 2 seconds. The relay, in turn, switches on the fuel pump. If the relay is bad, the pump won't work and the engine won't start.

It's worth a gamble to replace the relay if there is no whirring from the fuel pump when you turn on the ignition switch rather than to have the fuel tank removed and opened. The most difficult part of replacing the relay is identifying it. On many GM models, it is one of three relays mounted side by side on the firewall near the brake power booster. If you have a service manual for your car, use it to find the relay. If you don't, have a GM shop do the job to avoid replacing the wrong part.

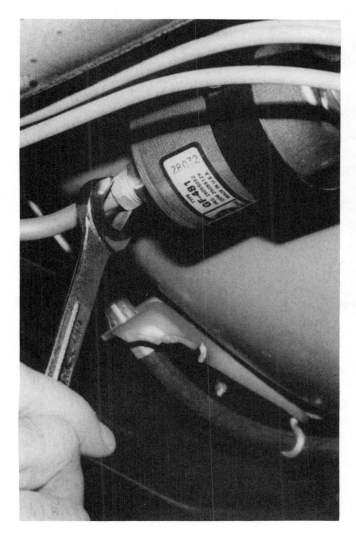

Fig. 7-6. Release fuel pressure before disconnecting the fuel line on each end of the fuel filter.

Fig. 7-7. To remove this filter, unscrew the bracket bolt (arrow), open the bracket, and slide the filter out of the bracket.

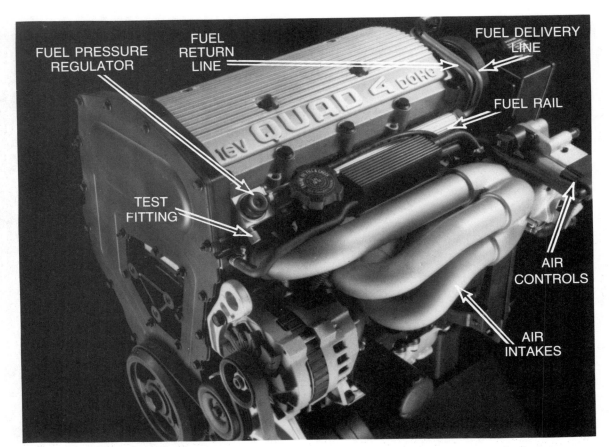

Fig. 7-8. The parts of the MPFI system used on the GM Quad 4 engine are typical of those on other GM engines having MPFI.

REASONS FOR HARD STARTING

Suspect one of three fuel delivery parts if you have to crank the engine for a long period before it will start: a bad fuel pump check valve, a faulty FPR, or a fuel injector that doesn't close.

Bad Check Valve

This valve, which is inside the fuel tank, is supposed to prevent gas from draining out of the fuel pump when the engine is turned off. If the valve malfunctions, the pump has to work harder to push gas to the injectors when the engine is started. This results in long cranking periods.

You can determine if the check valve is bad by connecting a pressure gauge to the fuel rail fitting as is done when doing a fuel injector balance test (Chapter 6). Turn on the ignition switch. Pressure should rise to between 234 and 325 kPa or 24 to 47 psi. Turn off the ignition switch. If pressure doesn't drop at once, but drops significantly within 15 minutes, the check valve is faulty. Remove the fuel tank to install a new valve.

Faulty Fuel Pressure Regulator

In the GM MPFI system, the FPR maintains a workable pressure at the fuel injectors in all kinds of conditions under which the engine must perform (Fig. 7-8). The FPR, which is mounted on the fuel rail, is a spring-loaded diaphragm that monitors fuel system pressure on one side and manifold (vacuum) pressure on the other (Fig. 7-9). There is a hose from it to the fuel tank. This is the fuel return hose. When the FPR senses that fuel pressure is getting too high, it opens to allow gas to return to the tank, thereby reducing pressure in the system to normal.

To test the FPR, do the following:

1. If pressure does not hold when doing a fuel pressure test, turn off the ignition switch for 10 seconds. Then turn it back on.

Fig. 7-9. The fuel pressure regulator, being pointed to, is mounted on the fuel rail.

2. As the fuel pressure gauge shows pressure hitting its peak again, pinch the fuel return hose closed before it has a chance to drop (Fig. 7-10). If pressure does not hold steady, you have a faulty FPR on your hands. Replace it.

Fuel Injector Does Not Close

A fuel injector that is stuck open causes excessive gas to flood the cylinder. Identifying an open injector is often easy to do.

Remove the engine's spark plugs after you have cranked the engine, but before it starts.

Fig. 7-10. The fuel pressure regulators of most GM engines having MPFI have a nipple to which the fuel return hose is attached. The fuel delivery line enters from the opposite side.

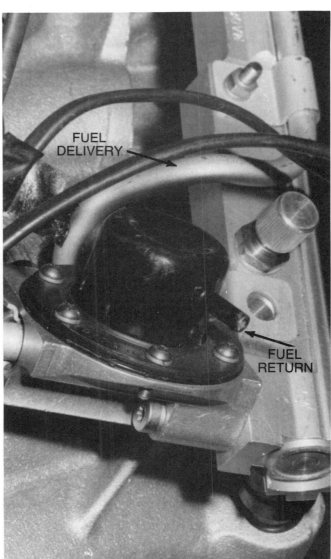

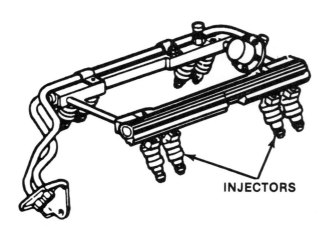

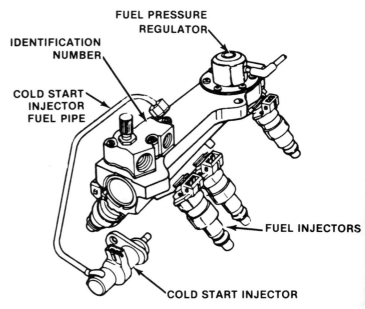

Fig. 7-11. Fuel injectors of a GM MPFI system are attached to the fuel rail. The assembly illustrated in this drawing is used on GM V-8 engines. To replace a faulty injector requires that you first remove the entire assembly from the engine.

If a plug is soaked with gas, the fuel injector serving that cylinder is stuck, resulting in flooding. Replace the faulty injector (Figs. 7-11, 7-12, 7-13).

Fig. 7-13. So-called first generation MPFI systems used by GM employ a cold-start system that features a cold-start injector that is controlled by a timer. The cold-start injector sprays gas into the engine. This shot coupled with gas being injected by the regular fuel injectors provides an engine with the extra gas it needs to start when it's cold. Later generation GM MPFI systems don't have a cold-start system. Instead, the ECM keeps the regular injectors open for a longer period when an engine is first started. Suspect a malfunction of the cold-start system if you have trouble starting the engine only when it's cold.

Fig. 7-12. The degree of difficulty in removing a fuel rail–fuel injector assembly from an engine depends on the type of engine. The degree of difficulty is related to the number of components that have to be removed to get at the fuel rail–fuel injector assembly.

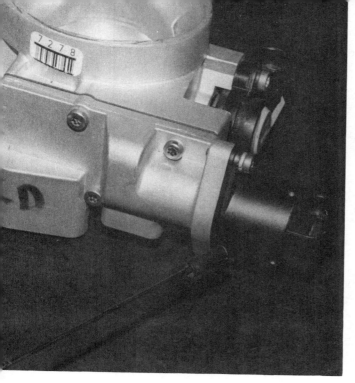

Fig. 7-14. The idle air control, being pointed to, is an electrically controlled unit attached to the air assembly housing. This housing is often referred to as the air plenum. It lies behind the throttle valve. Its job is to get air to the engine as the engine is idling and the throttle valve is closed.

UNDERSTANDING THE AIR CIRCUIT

If there is a malfunction in the air circuit, the result is rough idling. The trouble usually is concentrated in one of two parts: the air filter or the idle air control (IAC).

Replace the air filter, which may be dirty, and see what happens. If this doesn't help, turn to the IAC (Fig. 7-14). However, be aware that failure of the IAC to work may be caused by a glitch with the ECM, which controls it.

If the IAC stays in a fully open or partially open position when it's supposed to be closed, too much air will be introduced into the intake system and the engine will idle roughly (Figs. 7-15, 7-16). If the IAC gets stuck in the closed position when it's supposed to be open, idling speed will be too slow and the engine will stall.

Suppose one of these things is happening. Accurately testing the performance of the IAC requires grounding a diagnostic test terminal

Fig. 7-15. Before proceeding with tests to determine if the IAC is malfunctioning, examine the air duct. If clamps holding the duct at either end are loose, excessive air will flow into the air assembly housing and disrupt the way the engine idles.

Fig. 7-16. A split in the air duct will also cause too much air to get inside the engine. The result is rough idling.

Fig. 7-17. In addition to the air filter, air duct, and idle air control, a faulty throttle position sensor (TPS) will affect the way in which an engine with MPFI runs. A trouble code stored in the ECM and displayed on the CHECK ENGINE or SERVICE ENGINE SOON light on the dash when the ECM is activated will indicate whether the TPS should be replaced.

in the car. Explaining this is beyond the scope of this book, because the diagnostic test terminals and the codes that are displayed as a result vary from one GM model to another.

Is there a less complex method you can use at least to get an indication whether the IAC is performing properly? Yes, there is:

1. After allowing the engine to warm up, connect a tachometer.

2. With the transmission in Park or Neutral and the engine running at idle, note the rpm reading.

3. Turn off the engine for exactly 10 seconds. Start it again and immediately note the rpm. If there is no increase in rpm, you can stop right now. The problem you're having is being caused by a faulty IAC or ECM, or both.

4. If there is a rise in rpm, let the engine run for 1 minute more and again take a reading from the tachometer. If the rpm reading has returned to what it was the first time, there is nothing wrong with the IAC. Your problem is caused by something else. But if there is no return, the IAC or ECM, or both are in trouble.

Have a GM professional technician determine if the cause of the trouble is the IAC. The IAC is a much less expensive part than an ECM, and you'd feel awful after replacing an ECM to find out the IAC has been the faulty unit all along (Figs. 7-17, 7-18).

Fig. 7-18. The only other air-delivery part that can cause poor engine performance is the sensor of the mass air-flow meter. This sensor has been a particularly troublesome part. Keep this in mind if you or your mechanic have trouble pinpointing the reason for an engine-performance problem.

8 Ford Central Fuel Injection Systems

An Overview

Fig. 8-1. The Ford 3.8-liter V-6 engine is typical of an engine that uses the Central Fuel Injection (CFI) system.

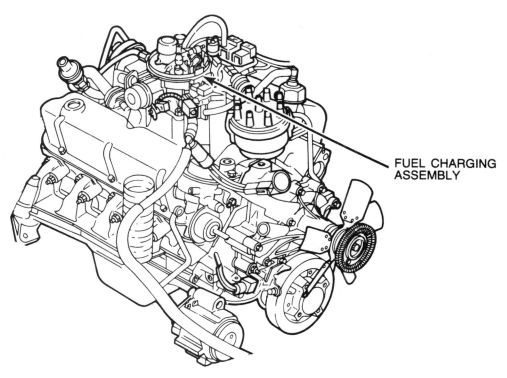

Fig. 8-2. The Ford CFI system has a fuel charging assembly (or throttle body) located on top of the intake manifold. If the unit you have looks like the one in this illustration, it possesses two fuel injectors and is a high-pressure system. This type of system is used on V-6 and V-8 engines.

THE EFI SYSTEMS used by Ford on a number of its engines are called central fuel injection (CFI) systems (Fig. 8-1). They are actually TBI systems similar to those used by GM and Chrysler. Whether called TBI or CFI, the systems are characterized by the placement of one or two fuel injectors in a throttle body (Ford calls it a fuel charging assembly) that is positioned atop the intake manifold where a carburetor would be if you were looking at an engine with a carburetor (Figs. 8-2, 8-3).

Fig. 8-3. A different CFI configuration is shown in this illustration. This fuel charging assembly (throttle body) is also located on top of the intake manifold. However, it has one fuel injector, is a low-pressure system, and is used on four-cylinder engines.

Fig. 8-4. The high-pressure CFI system is outfitted with the familiar-looking round air cleaner filter assembly, as are most carburetors.

Fig. 8-5. The low-pressure CFI system utilizes a rectangular air cleaner assembly. Air flows through the air cleaner and duct into the fuel charging assembly, past a throttle valve in the fuel charging assembly, and into the cylinders.

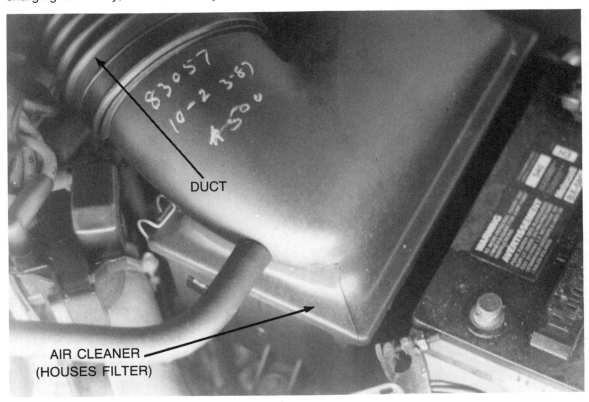

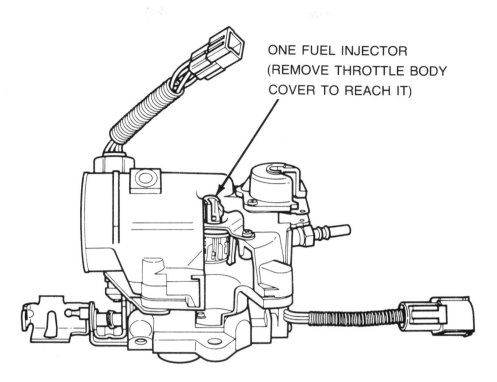

Fig. 8-6. This illustration shows the Ford low-pressure CFI system with one fuel injector.

Ford describes each of its CFI systems as a pulse–time-modulated injection system. There are two kinds: a low-pressure system and a high-pressure system (Figs. 8-4, 8-5). You can tell the difference between the two by counting the number of fuel injectors present in the fuel charging assembly. If there is one fuel injector, it is a low-pressure system (Fig. 8-6); if there are two fuel injectors, it is a high-pressure system (Fig. 8-7).

Both systems engage in two distinct operations: They take in air and they deliver gas. There is a third operation involved, but it is not performed by the CFI system. It is, however, closely associated to that system, because it controls the intake of air and the delivery of gas. That operation is conducted by the vehicle's electronic control system.

Fig. 8-7. This illustration shows the Ford high-pressure CFI system with two fuel injectors.

INTAKE OF AIR

The intake of air into the fuel charging assembly of a Ford CFI system is similar to the way air is taken into an engine having a carburetor. When the throttle plate is opened by the driver pressing the accelerator pedal, outside air, which is under normal atmospheric pressure (14.7 psi at sea level), is forced through the air intake of the fuel charging assembly. This takes place because pressure in the intake manifold is lower than atmospheric pressure.

Air flow into the engine of a high-pressure CFI system is down through the throttle body past the throttle plate into the intake manifold and cylinders (Fig. 8-8). There are two air horns, each equipped with a throttle plate. One air horn serves one bank of cylinders (remember, you're dealing with a V-6 or V-8 engine), while the other serves the other bank. Similarly, each of the two fuel injectors is aimed to spray gas into its respective throttle plate opening.

With a low-pressure system there is only one air horn (it is on the side of the fuel charging assembly), fuel injector, and throttle plate (Figs. 8-9 & 8-10).

Whichever system you deal with, keep in mind that the quantity of air that is permitted to flow into the engine is dependent upon how wide the throttle valve is opened as the driver presses on the accelerator pedal. This motion is transferred to the throttle valve by means of a throttle linkage, which is practically identical to that of an engine having a carburetor.

DELIVERY OF FUEL

Depending upon the system in the vehicle—low or high pressure—Ford utilizes either a high-pressure or low-pressure electric fuel pump. With the high-pressure system, a fuel pump positioned in a sump inside the fuel tank delivers gas into the fuel delivery line at a pressure of 39 psi. The pump is equipped with an internal relief valve that opens to allow

Fig. 8-8. This illustration shows the relationship of the air intake, fuel injector, and throttle plate on one side of the Ford high-pressure CFI system. Remember that there's an identical setup on the other side of the fuel charging assembly where the other fuel injector is positioned.

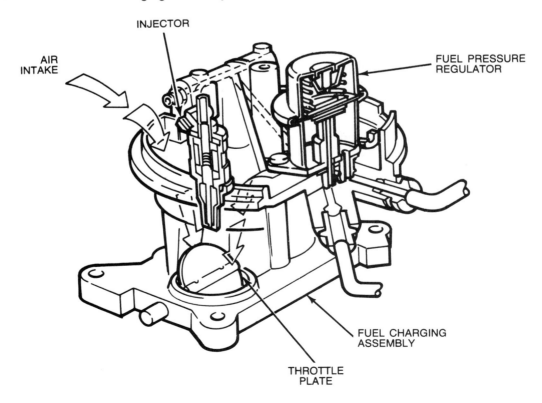

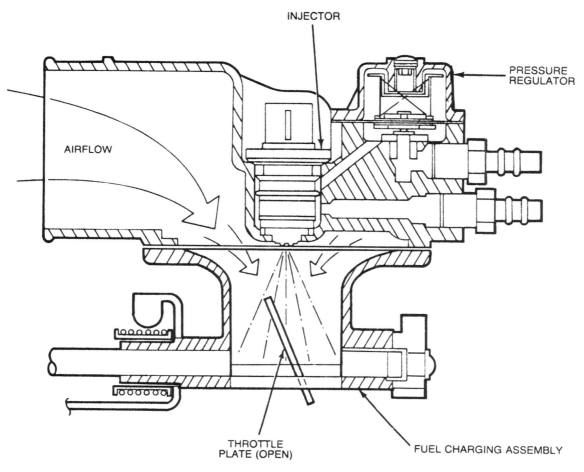

Fig. 8-9. This illustration shows the relationship of the single air intake, single fuel injector, and single throttle plate of the Ford low-pressure CFI system.

Fig. 8-10. This is the mouth of the air intake of the Ford low-pressure CFI system. Air coming from the air cleaner (Fig. 8-5) enters the mouth through an air duct.

Fig. 8-11. You will find two lines connected to the Ford CFI fuel charging assembly. The one having the larger diameter is the fuel-supply line. The one with the smaller diameter is the fuel-return line.

pressure to blow off if it builds up to a level in excess of 39 psi.

The pump used in the low-pressure system works like the high-pressure system pump, even to the extent of being equipped with overpressure protection. The only difference is that the low-pressure pump is designed to deliver gas at 14.5 psi.

In dealing with the fuel delivery system of a Ford CFI system, you will come across *two* fuel lines (Fig. 8-11). As you might expect, there is a line that delivers gas from the fuel tank to the fuel injector(s) in the fuel charging assembly. This line is the fuel supply line.

The fuel supply line possesses an in-line fuel filter through which gas flows on its way to the fuel charging assembly (Fig. 8-12). The purpose of the in-line filter is to trap dirt and moisture that might be carried by the gas. If dirt or moisture get to the fuel injector(s), it will soon foul the injector(s) and cause an engine performance problem.

The second fuel line is the fuel return line. It is connected to the fuel charging assembly and extends back to the fuel tank. Its purpose is to allow fuel not needed by the engine to return to the tank. A fuel pressure regulator (FPR) mounted on the fuel charging assembly

Fig. 8-12. This illustration shows the parts of a typical Ford vehicle utilizing CFI.

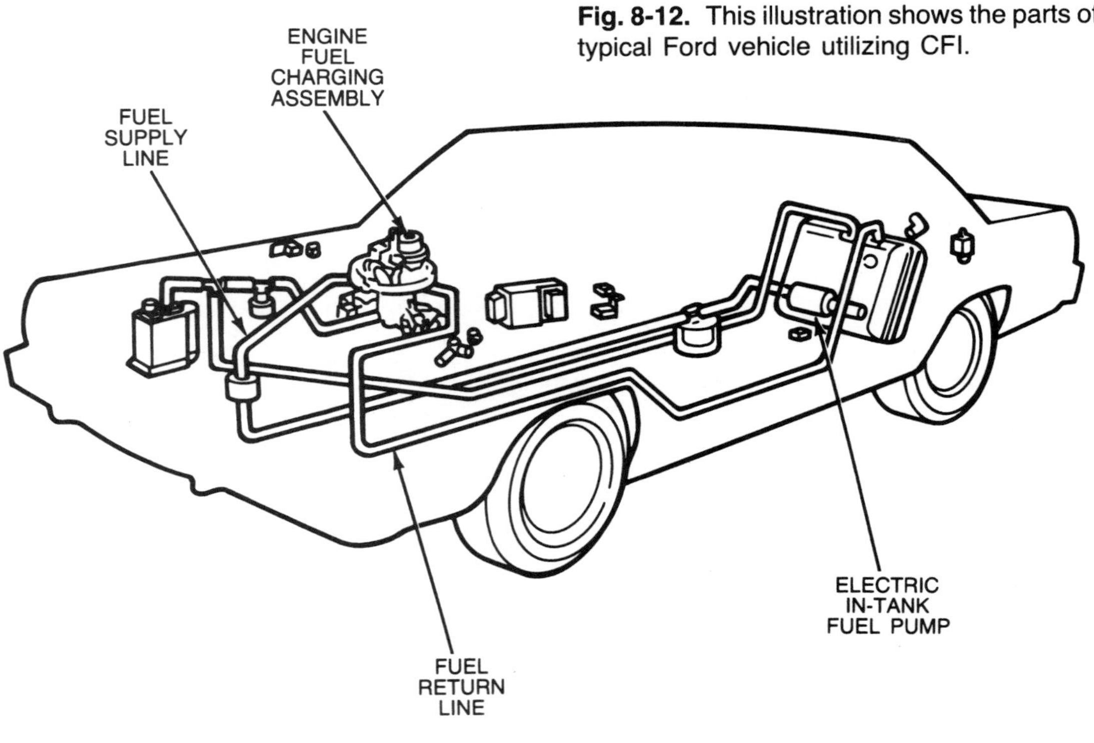

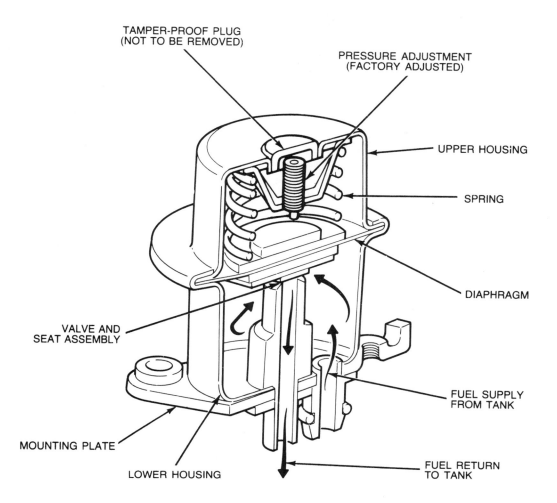

Fig. 8-13. The fuel pressure regulator is an important part of all Ford EFI systems—multipoint fuel injection (Chapter 10) as well as CFI.

is the key part in controlling just how much gas gets to the fuel injector(s) and how much returns to the fuel tank.

The FPR used with both the high- and low-pressure CFI systems is adjusted at the factory (Fig. 8-13). Its components are a spring and diaphragm assembly, which moves up and down with respect to pressure, a valve and seat assembly, and fuel supply and fuel return ports. The spring-loaded diaphragm frees the valve and seat assembly, thereby opening the fuel return port when the specified fuel pressure within the system is exceeded.

The FPR performs a task in addition to regulating fuel pressure. It traps fuel when the engine is shut down to prevent the formation of vapor in the fuel line. Vapor can cause a problem in restarting a warm engine.

In the high-pressure system, the FPR maintains fuel pressure within the system at 39 psi. It is attached to the fuel charging assembly and can be replaced if it fails.

In the low-pressure system, the FPR maintains fuel pressure within the system at 14.5 psi. The regulator is an integral part of the fuel charging assembly. If it fails, check with a dealer of Ford auto parts to determine if there is a repair kit available that will permit you to replace only the diaphragm. If there is not, you will have to replace the fuel metering cover that holds the FPR.

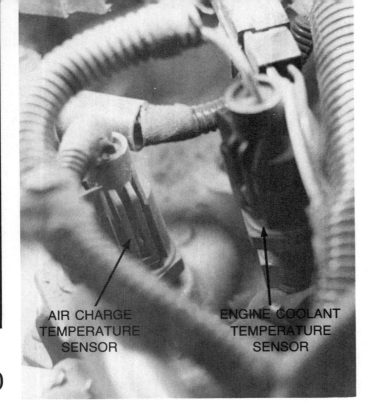

Fig. 8-14. A Ford engine equipped with CFI possesses a number of sensors that transmit data about the engine to a microprocessor. The microprocessor (or electronic control assembly) assimilates the data to determine the length of time the fuel injectors should remain open.

ELECTRONIC CONTROL SYSTEM

Electronic control of the Ford CFI system is accomplished by means of a microprocessor or electronic control assembly (frequently referred to as the ECA) that receives signals from a series of sensors scattered throughout the engine (Fig. 8-14).

The ECA makes computations based on the input signals from the sensors and transmits "commands" to the fuel injector(s). When things are working properly, these commands permit the maintenance of a precise air-to-gasoline ratio. If a sensor malfunctions, the ECA can keep the fuel injector(s) open or closed longer than necessary. The result is the delivery of an oversupply of gas or a lack of gas to the cylinders by the fuel injector(s), with the outcome being an engine performance problem. In other words, the fuel injection system may be working perfectly, but because of a defect in an electronic control unit the CFI may be prevented from performing properly.

To put the various players involved in Ford CFI system performance into perspective when an engine drivability problem arises, first test vacuum and electrical components (Chapter 1) (Fig. 8-15). Then, turn your attention to the mechanical parts of the CFI system. Finally, if this troubleshooting fails to reveal the reason for the performance letdown, look to the ECA.

Fig. 8-15. Don't overlook basics. Before tackling the CFI system to try to find a defect that is causing an engine-performance problem, remember that a cracked or loose vacuum hose or damaged vacuum component is more likely the reason. Test these parts as discussed in Chapter 1.

9

Ford Central Fuel Injection Systems

Troubleshooting and Repair

Fig. 9-1. The vacuum-operated positive crankcase ventilation (PCV) valve and hoses should be high on your list of things to check when an engine performance problem occurs.

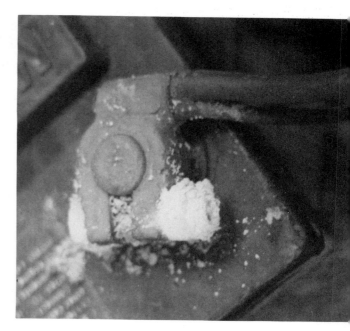

Fig. 9-2. Although the Ford CFI system is a modern computerized assembly, you shouldn't lose sight of basics. For example, battery cable connections can corrode. Corrosion can seriously disrupt engine performance. Therefore, examine cable connections often and clean them when necessary.

TROUBLESHOOTING AND REPAIRING a CFI system on a vehicle bearing the Ford, Mercury, or Lincoln name is easier than troubleshooting a carburetor. There are fewer parts involved. Furthermore, each part usually creates a well-defined performance problem when it malfunctions, making it a relatively simple matter in most instances to relate a performance problem to the part or parts causing it.

Information in this chapter will help you determine which part of a Ford CFI system is most likely causing a particular performance problem. To put matters into perspective, let's begin with a chart of malfunctions and their probable causes. Information you will need to troubleshoot the parts listed under "Probable Causes" is presented following the chart.

Note: Remember that most malfunctions listed in this chart are caused by vacuum loss and faulty electrical connectors (Figs. 9-1, 9-2). Refer to Chapter 1.

INOPERATIVE ELECTRIC BIMETAL THERMOSTAT

This unit is referred to in some Ford service manuals as a choke. You may have heard that engines having EFI systems do not have chokes. That's true—except for the Ford 5.0-liter engine with CFI (Fig. 9-3).

The purpose of the electric bimetal thermostat on the Ford 5.0-liter engine is to increase

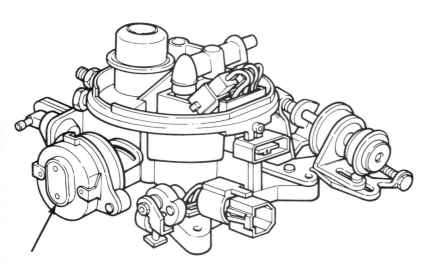

Fig. 9-3. Vehicles with 5.0-liter Ford engines and CFI have the part illustrated here (arrow). It resembles the choke housing you'll find on an engine having a carburetor and serves the same function.

Malfunction	Probable Causes
Hard starting	■ Inoperative electric bimetal thermostat ■ Leaking gasket between fuel charging assembly and intake manifold
Rough idling—cold engine	■ Inoperative electric bimetal thermostat ■ Loose or cracked air cleaner duct ■ Incorrect idling speed
Stall, stumble, hesitation	■ Inoperative electric bimetal thermostat ■ Incorrect idling speed ■ Clogged fuel filter ■ Weak fuel pump ■ Dirty air cleaner filter
Stalling as engine decelerates or on quick stops	■ Leaking gasket between fuel charging assembly and intake manifold ■ Inoperative throttle positioner
Lack of power	■ Sticking throttle linkage ■ Clogged fuel filter ■ Bad fuel injector ■ Damaged FPR ■ Weak fuel pump
Surging	■ Clogged fuel filter ■ Weak fuel pump

engine idling speed when the engine is cold in order to avoid hard starting, rough idling, stalling, stumbling, and hesitation on acceleration. A cam controlled by the bimetal thermostat heating element is positioned to provide the fast idling speed.

When the engine is cold, the bimetal element, which is a coil, is wound tightly to keep the cam positioned on the high step of the throttle lever. This keeps the throttle valve open wider than normal, which results in a faster idling speed. When the engine is warmed up and a faster idling speed is no longer necessary, the bimetal coil becomes hot and unwinds. As it unwinds, it releases the cam from the throttle lever. Engine idling speed assumes a slower, more normal pace.

When a problem arises that might involve the choke, the first step to take is to make sure the external parts of the mechanism are free and not binding. Spray the mechanisms with a choke cleaner and wipe them clean.

If this step fails to resolve the performance problem, determine whether the electric heating element is working properly. Pull the electric wire off the choke housing and connect an ammeter between the terminal on the housing and the wire. Start the engine. The ammeter should show a surge of current. Then, as the engine runs at idle, the ammeter should register 2 amps or less for 20 seconds. If at the end of 60 seconds the ammeter is still showing 1 amp or more, replace the thermostat housing.

Suppose the ammeter shows that current is not reaching the choke housing and, consequently, the bimetal coil. Suspect a faulty wire that feeds electricity from the battery to the choke housing. That wire may have to be replaced.

LEAKING GASKET BETWEEN FUEL CHARGING ASSEMBLY AND INTAKE MANIFOLD

Use a vacuum gauge to determine whether the gasket between the fuel charging assembly and intake manifold has failed (Fig. 9-4). A bad gasket allows excess air into the fuel intake system, causing a lean fuel mixture, which can make an engine hard to start or cause it to stall once it does start.

With the engine cold, attach the vacuum gauge to a vacuum fitting on the fuel charging assembly or on the intake manifold. Start the engine and let it run at idle for a few minutes. Then, check the vacuum gauge needle. If the needle is steady (no flutter), but shows a vacuum reading lower than that regarded as normal, bolts holding the fuel charging assembly to the intake manifold may be loose (tighten them and test again) or the gasket

Fig. 9-4. A vacuum gauge, either as an independent tool or as part of a hand-vacuum pump, is indispensable for tracking down components that are losing vacuum.

Fig. 9-5. This photograph illustrates the air cleaner cover and duct of a Ford low-pressure CFI system; however, servicing a high-pressure CFI system is done similarly. Make sure the filter is clean, the duct is not cracked, and clamps are tight.

between the fuel charging assembly and intake manifold may have failed.

A normal vacuum reading varies from engine to engine, so check the service manual. Generally, though, a reading of 17 psi or higher is considered within normal limits for Ford Motor Company engines with CFI.

To verify if a low vacuum gauge reading is the result of a bad gasket, connect a tachometer to the engine and allow the engine to run at idle speed. Fill an oilcan possessing a long snout with SAE 30 oil and apply oil to the joint where the fuel charging assembly attaches to the intake manifold. If the tachometer shows a temporary increase in idle speed, there is a leak. Replace the gasket.

LOOSE OR CRACKED AIR CLEANER DUCT

Tighten the clamps that hold the duct to the air cleaner. Then, inspect inside the folds to see if the duct is torn, in which case you should replace the duct (Fig. 9-5).

As long as you are working at the air cleaner, inspect the air cleaner filter on the chance that a dirty filter is resulting in a rich fuel condition. A mixture that is too rich will cause a performance problem as quickly with CFI as it does with a carburetor fuel system. Therefore, inspect for a dirty air cleaner filter at least once every 15,000 miles.

INCORRECT IDLING SPEED

You may be able to adjust the curb (also called slow) and fast idling speeds. The following data provides examples of what can and can't be done to some Ford engines with CFI. For information about other engines refer to applicable service manuals.

3.8-Liter Engine

The curb idling speed of this engine is controlled by the engine electronic control system microprocessor. Thus, if a drivability problem arises that might be caused by a higher than normal curb idling speed and the procedure described here does not correct the problem, test the engine electronic control system.

Check the curb idling speed of a 3.8-liter engine equipped with CFI as follows:

1. Set the parking brake, block the wheels, warm up the engine, make sure all accessories are turned off, and connect a tachometer.

2. Allow the engine to idle for 60 seconds with the transmission in Drive or Reverse. After 60 seconds the tachometer should record a curb idling speed that is in line with the specification printed on the vehicle emis-

Fig. 9-6. The part illustrated in this photograph is an idle speed control (ISC) on a 3.8-liter Ford engine with a CFI system. The plunger of the ISC comes into contact with the throttle lever to maintain the engine at a specified speed when it is idling.

Fig. 9-7. This drawing provides perspective that will help you identify the ISC in respect to its position on the fuel charging assembly of a 3.8-liter engine.

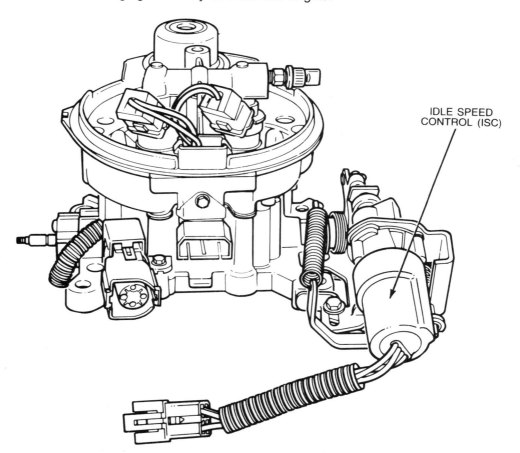

sion control information (VECI) decal. The VECI decal is mounted in the engine compartment.

3. If the curb idling speed exceeds the specification printed on the decal, see if the throttle lever is touching the idle speed control (ISC) motor (Figs. 9-6, 9-7). If it isn't, shut off the engine, take off the air cleaner, and disconnect the throttle lever from the throttle stop adjustment screw in this way:

- In the engine compartment, find the self-test connector and self-test input connector. These two connectors are next to each other.

- Attach a jumper wire between the self-test input connector and the signal return pin of the self-test connector.

- Turn the ignition key to the Run position, but do not start the engine. Check to see if the plunger of the ISC motor has retracted. If not, test the engine electronic control system as outlined in the service manual.

- If the plunger of the ISC motor has retracted, wait 10 seconds, turn off the ignition switch, and remove the jumper wire. Now, turn the throttle stop adjustment screw until it comes loose, throw it away, and install a new screw (Fig. 9-8). The correct screw can be obtained from the parts department of a Ford dealer.

- With the throttle plate closed, turn the new throttle stop adjustment screw counterclockwise until a gap of 0.005 inch exists between the tip of the screw and the throttle lever. Now, turn the screw an additional 1½ turns.

Fig. 9-8. You have to work with the throttle stop adjustment screw to attain a proper idling speed from a 3.8-liter Ford engine having CFI.

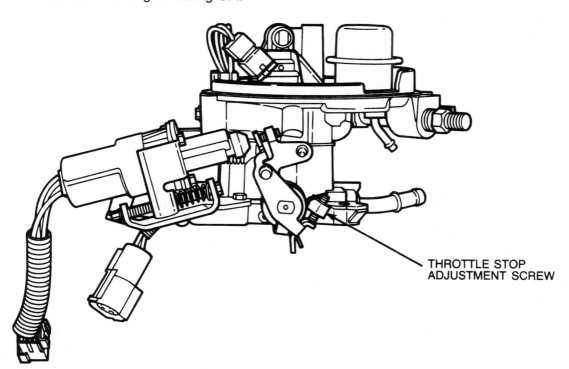

- You now have to set the fast idle. Remove the rubber duct cover from the tip of the ISC motor and loosen the lock screw holding the tip. Push the tip in toward the ISC motor and position the shank of a 9/32-inch drill bit between the tip of the motor and the throttle lever. Retighten the lock screw, reinstall the dust cover, and reattach the air cleaner.

5-Liter Engine

To check and adjust the curb idling speed of a 5.0-liter engine with CFI, proceed as follows:

1. Connect a tachometer, put the transmission control lever into Neutral, engage the parking brake, block the wheels, and warm up the engine. Place the air conditioner/heater selector lever in the Off setting and turn off all other accessories.

2. For all models except Mustang and Capri, shut off the engine and restart it at once. Allow it to run at a fast idling speed of 2,000 rpm for 60 seconds; then let it return to slow idling speed for 30 seconds. For Mustang and Capri, do not shut the engine off, but boost the idling speed to 2,000 rpm for 15 seconds. Then let it return to slow idling speed for at least 10 seconds more.

3. Shift the transmission into Reverse and check the curb idling speed. It should be at the setting shown on the VECI decal. If this is not the case, take these steps:

- Loosen the saddle bracket locking screw on the antistall dashpot (Fig. 9-9).

- If curb idling speed is below that specified on the VECI decal and the car is not a Mustang or Capri, shut off the engine and turn the saddle bracket adjusting screw 1 full turn clockwise. Restart the engine and run it at a fast idling speed of 2,000 rpm for 60 seconds. Allow it to return to curb idling

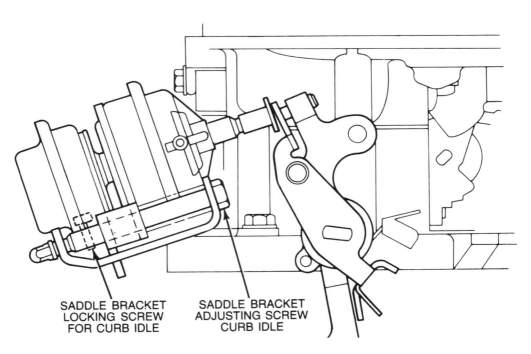

Fig. 9-9. The CFI system of a Ford 5.0-liter engine allows you to adjust the slow (curb) idle speed. This drawing points out the saddle bracket locking screw and saddle bracket adjusting screw of the antistall dashpot, which are used in adjusting curb idle.

Fig. 9-10. The Ford CFI system is one of the few TBI systems that provides a diagnostic pressure test valve (arrow).

- For all models with 5.0-liter engines and CFI that show a curb idling speed above that specified on the VECI decal, loosen and turn the saddle bracket adjusting screw counterclockwise until the specified idling speed is reached.

Important: After making the curb idling speed adjustment, be sure to tighten the saddle bracket locking screw.

speed for 30 seconds. Continue to do this until the engine begins running at the curb idling speed setting specified on the VECI decal, at which time tighten the saddle bracket locking screw.

- If curb idling speed is below that specified on the VECI decal and the car is a Mustang or Capri, do not shut off the engine. Turn the saddle bracket adjusting screw until the engine begins to run at the curb idling speed is played on the VECI decal.

SERVICING THE FUEL SYSTEM

This section of the chapter discusses how to proceed when it becomes necessary to check the parts of the CFI fuel delivery system that might be the cause of one of the conditions outlined in the troubleshooting chart. These parts are the fuel pump, fuel filter, FPR, and fuel injector(s). Each is described below, but before tackling any of them be aware that there is a diagnostic pressure test valve on the throttle body (Fig. 9-10). This valve allows you to tap a pressure gauge into the system to find out if the system is maintaining pressure (Fig. 9-11).

Fig. 9-11. To use the valve for the purposes discussed in the text, unscrew the cap. When you are done, be sure to replace the cap to prevent damage to the valve.

Fig. 9-12. Access to the fuel pump shut-off (inertia) switch is in the luggage compartment. To reactivate the switch, stick a finger through the hole and press the reset button. If your car does not have an access hole, the panel has to be removed to reach the switch.

The diagnostic pressure test valve also lets you purge the system of pressure—a must before disconnecting fuel line fittings. If you don't release pressure, gasoline will spray all over the place when you loosen a fuel line fitting to service the system. In addition, after servicing the system, you should use the test valve to purge air that may be introduced into the system as it is being serviced.

The diagnostic pressure test valve is a Schrader valve, which is similar to the air valve on a tire. You need a special tool to test the fuel delivery system using the diagnostic pressure test valve and also to release pressure. This tool, which can be purchased from a Ford/Mercury/Lincoln dealer or from an auto parts supply dealer, is called the Ford EFI Pressure Gauge and connects directly to the diagnostic pressure test valve. It bears Rotunda part number T8OL-9974-A. Do not attempt to work on a Ford CFI system using any other gauge.

Important: When you release fuel system pressure, make a repair, and then allow pressure to build back up by turning on the ignition, you must again release pressure if you are going to make additional repairs.

TESTING THE FUEL SYSTEM

A malfunction in the fuel delivery system results in a drop in pressure and a reduced flow of gasoline or no flow at all to the fuel injector(s). Obviously, if there is no flow, the engine won't start. To determine if the cause of an engine performance problem is in this system, do the following:

1. Make sure there's gas in the fuel tank.

2. Look for a gas leak at each fuel system fitting; then examine fuel lines to make sure none is cracked or kinked.

3. Disconnect the electrical connector lying just outside the fuel tank where the electric feed wire enters the tank. In the tank, this wire connects to the electric fuel pump. Attach a voltmeter to the connector that is part of the electric feed wire coming from the engine compartment. Have the ignition key turned on as you watch the voltmeter. The voltmeter needle should rise to about 12 volts, stay there for about 1 second, and drop to 0. This verifies that voltage from the battery is being delivered to the fuel pump. If the voltmeter does not respond this way, check the fuel pump inertia switch.

The purpose of the inertia switch is to shut off the fuel pump circuit automatically if the vehicle is in an accident. This safety feature prevents gasoline from spraying through a ruptured fuel line and causing a fire. Find the inertia switch in the luggage compartment or in the spare tire well (Fig. 9-12).

The inertia switch may have failed, and this is the reason for the lack of electricity needed for the fuel pump to operate. To check this, press the reset button on the inertia switch and do the voltmeter test again. If this results in a normal voltmeter reading, but the result is temporary, replace the inertia switch.

If the inertia switch is not the reason for the faulty voltmeter reading, proceed with troubleshooting.

4. Turn off the ignition switch and connect an ohmmeter to the connector of that part of the electric feed wire that enters the fuel tank (Fig. 9-13). If there is a lack of continuity, have the fuel tank taken from the car and the fuel pump removed from the tank. A continuity test should then be made at the electric terminals on the pump to which the feed wire attaches. If there is continuity, the electric feed wire is faulty and should be replaced. If there still is a lack of continuity, replace the fuel pump.

5. Locate the fuel return line at the fuel charging assembly. The fuel return line is the line extending from the fuel charging assembly to the fuel tank. Disconnect that line at the fuel charging assembly.

Caution: Wipe up any gasoline spills.

Attach a four-foot hose to the fuel return line connection on the fuel charging assembly. Lay the other end of the hose in a 1-quart glass measuring cup.

Now, connect a CFI pressure gauge to the fuel diagnostic valve. Make sure the connector attaching the electric feed wire to the fuel pump wire discussed in step 3 is disconnected. As mentioned, this connector lies just outside the gas tank. Attach a jumper wire to the connector of the wire that enters the gas tank. This jumper

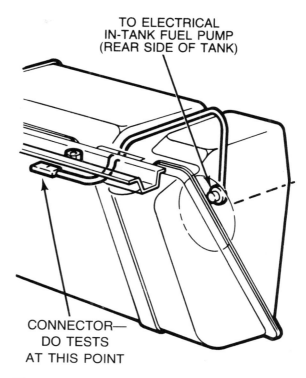

Fig. 9-13. Testing the CFI system requires working at the gas tank with the electrical connector that serves the fuel pump. That connector is positioned just outside the gas tank.

should be long enough to reach the battery.

Turn on the ignition switch and hold the other end of the jumper wire to the positive terminal of the battery for exactly 10 seconds. Then pull it away. This action will energize the fuel pump and allow a certain amount of fuel to flow into the measuring cup.

Record the pressure gauge reading and note the amount of fuel in the measuring cup. Here's how to interpret your findings:

■ If the fuel pressure gauge records 35 to 45 psi and maintains a reading of at least 30 psi immediately after you pull the jumper wire away from the battery, and if you get at least 7.5 ounces of fuel in the measuring cup, the CFI system is operating normally.

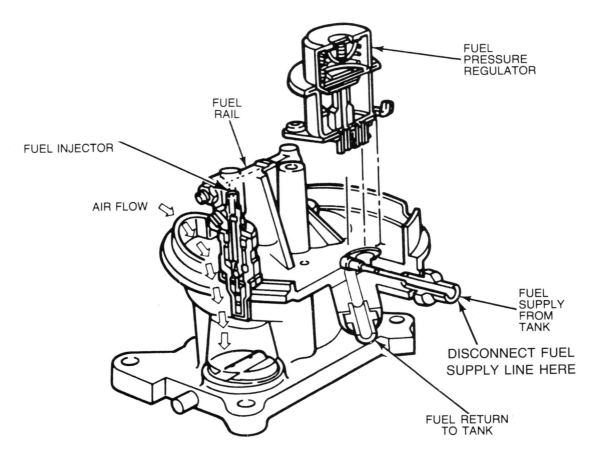

Fig. 9-14. If a normal supply of gas isn't getting to the fuel charging assembly because the fuel-supply line is clogged, the engine will falter. To make sure this is not happening, disconnect the line and use compressed air to blow out any debris.

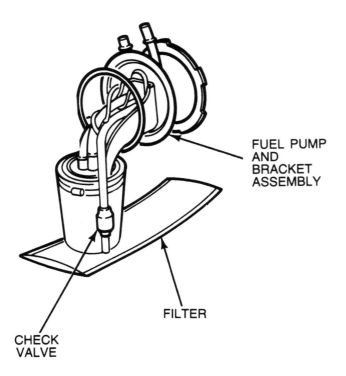

Fig. 9-15. The fuel pump used in a Ford CFI system incorporates a filter (sock), check valve, and fuel pump.

■ If the fuel pressure gauge records 35 to 45 psi and maintains a reading of at least 30 psi immediately after you pull the jumper wire away from the battery, but you don't get at least 7.5 ounces of gasoline in the measuring cup, there is a blocked fuel filter or fuel delivery line, or the fuel pump is not delivering enough gas. If replacing the fuel filter and blowing compressed air through the fuel delivery line don't resolve the condition, replace the fuel pump (Figs. 9-14, 9-15).

■ If the fuel pressure gauge records 35 to 45 psi and at least 7.5 ounces of gas are delivered, but a pressure reading of at least 30 psi is not maintained when you de-energize the fuel pump, the FPR or fuel injector(s) is leaking. Replace the fuel pressure regulator. If that doesn't resolve

Fig. 9-16. If to this point you have not found the reason for trouble, it is probably being caused by a faulty fuel pressure regulator or fuel injector(s). Replace the fuel pressure regulator first. The hose covering the air cleaner retainer was put in place purposely to prevent damage to the retainer while making repairs. If you take this precaution, remember to remove the hose before installing the air cleaner.

the matter, replace the injector(s) (Figs. 9-16, 9-17).

- If there is no flow of gasoline or no pressure (the engine won't start), replace the fuel pump and the coarse-mesh filter (or sock) that is part of the fuel pump assembly.

Fig. 9-17. The final step to Ford CFI troubleshooting is replacing the fuel injector(s). This is done by disconnecting wires and unscrewing the fuel rail to gain access to the injectors, which can then be popped from their seats. The task is not complicated.

10

Ford Multipoint Fuel Injection Systems

Troubleshooting and Repair

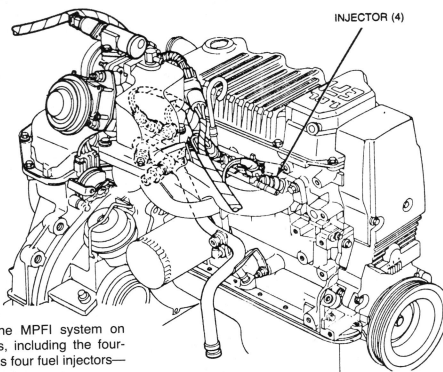

Fig. 10-1. Ford uses the MPFI system on several different engines, including the four-cylinder model, which has four fuel injectors—1 for each cylinder.

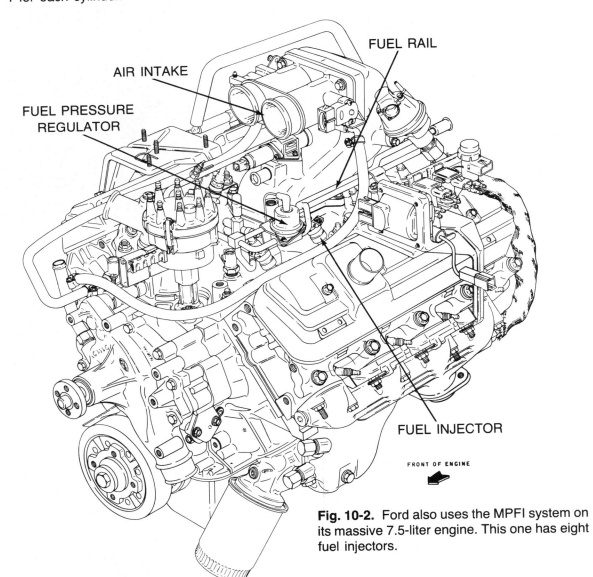

Fig. 10-2. Ford also uses the MPFI system on its massive 7.5-liter engine. This one has eight fuel injectors.

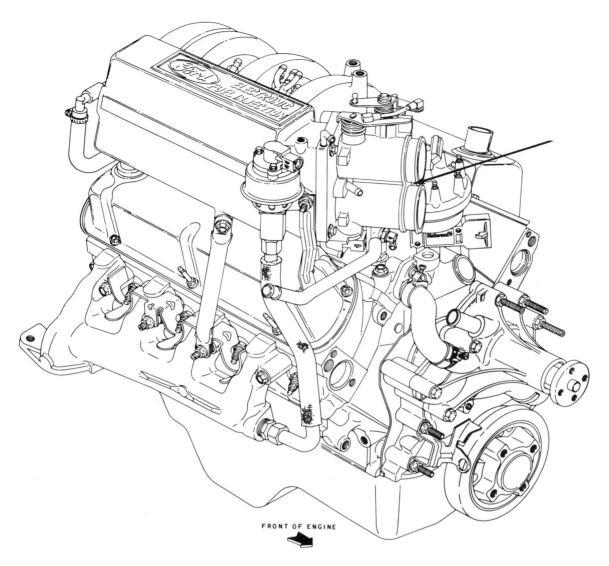

Fig. 10-3. The air intake unit of a Ford engine equipped with MPFI is shown in this illustration. Air is sent through a duct to the air horns by way of an air intake assembly, which includes an air filter.

THE FORD MULTIPOINT FUEL INJECTION (MPFI) system utilizes one fuel injector per cylinder. Each fuel injector is positioned to inject gas directly into a cylinder through the intake valve. Thus, an engine with MPFI will have four, six, or eight fuel injectors, depending on whether it is a four-, six-, or eight-cylinder engine (Figs. 10-1, 10-2). Simultaneously with the injection of gas, air is drawn into the engine through an air intake unit, ending up at each intake valve, where it mixes with gas (Fig. 10-3).

The length of time each fuel injector is kept open is controlled by the engine electronic control system. That length of time depends upon the conditions under which the engine is operating. Various sensors monitor those conditions and relay the data to the computer.

If a sensor malfunctions, the flow of gas to the cylinders is disrupted and one or more

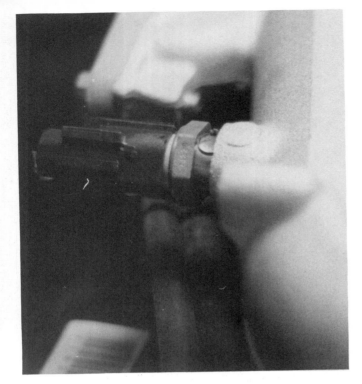

Fig. 10-4. Ford engines equipped with the MPFI system use a variety of sensors, such as this air charge temperature sensor, to monitor performance factors. Other sensors keep tabs on coolant temperature, the oxygen content of exhaust gas, vehicle speed, and manifold pressure.

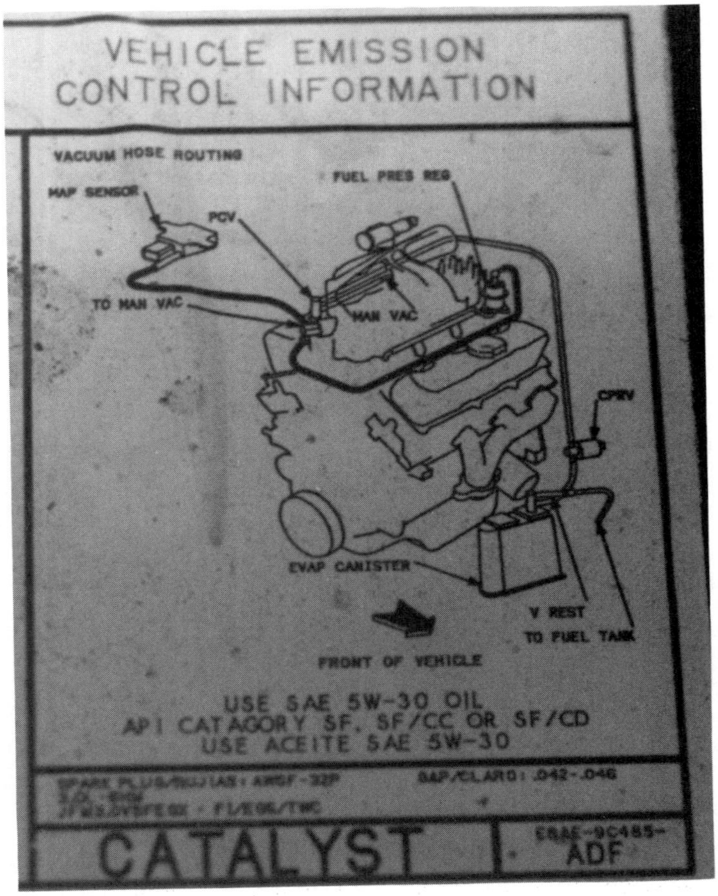

performance problems result, including hard starting, stalling, hesitation, rough idling, dieseling (engine run-on), lack of power, and engine stumble (Fig. 10-4). Therefore, if troubleshooting the fuel injection system as described below fails to reveal the reason for a problem, and vacuum loss and faulty electrical connections are not to blame (Chapter 1), a likely possibility is a faulty sensor (Fig. 10-5).

GETTING READY TO TROUBLESHOOT THE FORD MPFI SYSTEM

Suppose your Ford, Mercury, or Lincoln is experiencing one of the problems just mentioned. Before getting into troubleshooting, determine if the system uses one or two fuel pumps. Raise the car and examine the area just outside the fuel tank. Proceed forward toward the engine compartment. If you come upon a fuel pump, your vehicle has two fuel pumps—a low-pressure one and a high-pressure one. The low-pressure pump sits in a cavity in the fuel tank and sends gas to the high-pressure pump. The high-pressure pump, which is externally mounted, delivers the gas to the fuel injectors.

In appearance, both pumps are the same. However, a resistor wired into the electrical circuit of the low-pressure pump reduces its operating voltage to 11 volts. This causes the pump to deliver gas at a rate of approximately 14.5 psi. The high-pressure pump, on the other hand, doesn't have a resistor and operates at about 39 psi.

If you find no external fuel pump, your car has one pump. It is in a cavity inside the fuel tank and delivers gas at a rate of 39 psi (Fig. 10-6).

Fig. 10-5. Before troubleshooting the MPFI system, make sure a faulty vacuum component is not the reason for an engine performance trouble. In the engine compartment of a Ford vehicle, you will find a vacuum hose routing diagram similar to this one, which will identify vacuum components and their hoses.

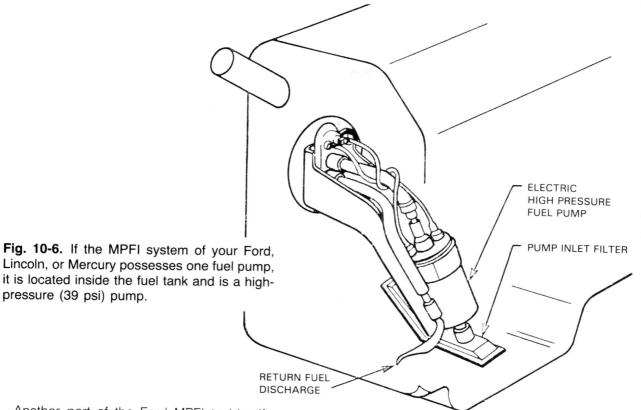

Fig. 10-6. If the MPFI system of your Ford, Lincoln, or Mercury possesses one fuel pump, it is located inside the fuel tank and is a high-pressure (39 psi) pump.

Another part of the Ford MPFI to identify before you proceed with troubleshooting is the diagnostic pressure valve. The diagnostic pressure valve is on the fuel rail—the elliptical fuel line circumventing the engine through which gas is delivered to each fuel injector. It resembles the valve of a tire with a removable cap over it and is used for connecting a pressure tester. The diagnostic pressure valve is also used to release pressure from the fuel system before replacing parts, to avoid having gas spray all over the place and all over you.

THE TROUBLESHOOTING PROCEDURE

To determine whether the cause of the engine performance problem the vehicle is experiencing lies with the MPFI system, proceed with troubleshooting as follows:

1. Make sure there is an adequate supply of gas in the fuel tank.

2. Examine the fuel system for leaks. Begin at the fuel tank and work forward toward the fuel injectors. If a leak is apparent at a fitting, tighten the fitting. If there is a leak from a fuel line, replace the line (Fig. 10-7).

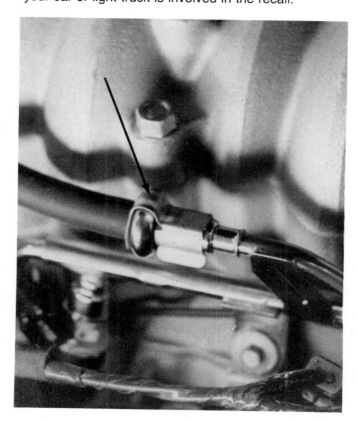

Fig. 10-7. Ford has announced a recall of over one million cars and light trucks to install a positive locking device (arrow) at the junction point of the steel fuel delivery line and fuel delivery hose. If your vehicle does not possess this device, check with a dealer to determine if your car or light truck is involved in the recall.

3. Assuming the car has one fuel pump, disconnect the electrical connector lying just outside the fuel tank. This separates the two parts of the wire carrying electricity to the fuel pump in the tank. Now, attach a voltmeter to the connector of the part of the wire coming from the engine compartment. Have someone in the car turn the ignition key on (do *not* crank the engine) as you watch the voltmeter. The voltmeter should show a reading of 10.5 to 12 volts. It should hold steady at that reading for about 1 second, and then fall to 0. If this does not happen, test the operation of the inertia switch (Chapter 9). If doing that doesn't uncover why there is an electrical fault, have the car's electrical system tested.

   Now, turn off the ignition switch, disconnect the voltmeter, and connect an ohmmeter to the part of the electrical connector attached to the wire that enters the fuel tank and attaches to the fuel pump. Check for continuity. If there is no continuity, remove the fuel pump from the fuel tank and test it on a bench by attaching the ohmmeter to the fuel pump terminals. Now if there is continuity, the wire is bad. But if there is still no continuity, the fuel pump is bad.

4. If the car has two fuel pumps, there is an electrical connector at the high-pressure (external) pump. Pull this connector apart and test voltage as was described for the single pump set up in step 3. Then test the continuity of the high-pressure pump by following the instructions outlined in that step.

   The differences in troubleshooting a dual- and single-pump arrangement lie with using the ohmmeter to test the low-pressure pump (in the fuel tank) of a vehicle with two pumps. This is done by connecting the ohmmeter across the terminals of the disconnected connector. The ohmmeter should show about 5 ohms. If this is not the case, remove the fuel pump from the fuel tank to test continuity at the fuel pump terminals, thus establishing whether the wire attached to the terminals is faulty, or the reason for your drivability problem lies with the fuel pump itself.

5. After testing electrical supply to the fuel pump(s), proceed to check the overall operation of the fuel delivery system in this way:

■ Find the fuel pressure regulator (FPR) and identify the fuel return line (Fig. 10-8). This

Fig. 10-8. This photograph shows what the fuel pressure regulator of a Ford MPFI system looks like. The line connected to the top of the regulator (arrow) is a vacuum reference line through which data relative to fuel pressure is transmitted to the engine computer.

is the line through which gas not needed to maintain pressure in the system is returned to the fuel pump. The fuel return line extends from the fuel FPR to the fuel tank. Disconnect this line from the FPR.

Caution: If any gas drips, wipe it up.

- Attach a 4- or 5-foot length of hose having the same inner diameter as the fuel return line to the FPR. Place the other end in a glass container that holds at least 1 quart. The container should be calibrated in ounces.

- Remove the cap from the diagnostic pressure valve on the fuel rail and attach the fuel pressure gauge to the valve. (*Note:* Ford recommends using the Ford EFI pressure gauge bearing Rotunda part number T80L-9974-A.) Again, if a little gas spurts out, wipe it up before proceeding.

- Pull apart the fuel pump wire connector (see steps 3 and 4 above). Connect one end of a long jumper wire equipped with alligator clips to the part of the connector attached to the wire extending to the fuel pump—not the wire coming from the engine. Make sure there is good metal-to-metal contact between the clip on the end of the jumper wire and the terminal inside the connector. Extend the other end of the jumper wire to the battery, but do not connect it.

- Turn the ignition switch on and hold the clip on the end of the jumper wire to the *positive* terminal of the battery for exactly 10 seconds. Check the pressure gauge as you do this. If the gauge doesn't show any pressure, make sure all connections are secure and that you're holding the jumper against the *positive* terminal of the battery. As the battery energizes the fuel pump(s), gas should flow into the glass container. At the end of 10 seconds, note the results. They should be all of the following:
  a. A pressure reading of 35 to 45 psi.
  b. If you have a single pump, no less than 7.5 ounces of gas delivered into the glass container in 10 seconds. If it is a dual-pump arrangement, no less than 9.5 ounces of gas in 10 seconds.
  c. After you pull away the end of the jumper wire from the battery, a pressure reading of at least 30 psi for a brief period of time.

If these conditions are met, you have no problem with the MPFI system. Look elsewhere for the cause of the performance problem. However, if one or more of the conditions are not met, here's what it means:

1. If the pressure reading is to specification, but the flow of gas isn't adequate, there is a clogged fuel filter or blocked fuel line.

2. If both pressure and gas flow meet the criteria, but a pressure of at least 30 psi is not maintained after you pull the jumper wire away, there is probably a malfunction with the FPR or one or more of the fuel injectors.

3. If there is no gas flow and there is no pressure (the engine won't start), the fuel pump(s) is faulty. Replace one or both, and also replace the fuel sock on the end of the fuel pickup pipe in the fuel tank.

The remainder of this chapter discusses replacing suspected faulty parts of the Ford MPFI delivery system. The examples are provided to give you an idea of what's involved, so you can decide whether you want to do the work yourself or turn it over to a service technician. If you decide to tackle the job, you will need the specific procedure for the engine you're working on, which is available in the relevant shop manual.

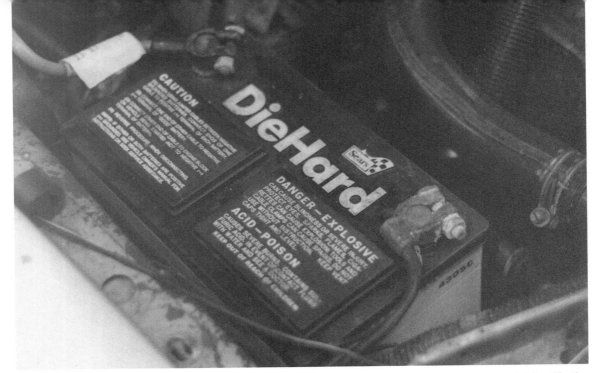

Fig. 10-9. Disconnect the battery negative cable before you begin to work. If you can't identify the negative cable, disconnect both cables.

Caution: Before you begin working, disconnect the cable from the negative terminal of the battery (Fig. 10-9) and relieve fuel system pressure by connecting the fuel pressure gauge to the diagnostic pressure valve and opening the bleed valve on the gauge. Remember that you are dealing with a system that carries gas. Work in a well-ventilated area, do not smoke or bring a flame near the car, and keep a fire extinguisher handy.

REPLACING THE FUEL FILTER

Before it enters the engine of a Ford, Mercury, or Lincoln equipped with MPFI, gas goes through a number of filters. Starting in the fuel tank, the first of these is a fuel pump inlet filter often referred to as a sock. It is a nylon element that is mounted on the fuel pump assembly. Its job is to make sure no contamination gets inside the fuel pump. If this filter clogs, which is a remote possibility since the filter is made of a coarse mesh that is cleaned as gas washes over it, gas will not get through to the fuel pump. Consequently, gas won't get to the engine. Replacing this filter is best left to a professional mechanic who has special equipment to remove gas from the tank safely in preparation for removing the tank from the vehicle to extract the fuel pump assembly.

The fuel filter, which you probably can replace yourself, is an in-line filter that traps small particles that may get through the sock and into the fuel delivery line. On cars equipped with one fuel pump, the in-line filter is located just outside the gas tank. Look for it on the underbody of the vehicle, forward of the right or left rear wheel well. If your car has two fuel pumps, the in-line filter is probably part of the high-pressure (external) fuel pump assembly. The filter and high-pressure pump are mounted on a bracket and are attached to one another by a fuel line. Replacing either type of filter is done as follows:

1. Remove the fuel tank cap.

2. Using the Rotunda T80L-9974-A fuel pressure gauge, depressurize the fuel system. Be sure to wipe up any gas that spills.

3. Remove the fuel lines at each end of the fuel filter. The lines may be held by quick-disconnect fittings, which are snapped open.

4. Undo the retaining bolt(s) and pop the fuel filter and retainer from the bracket.

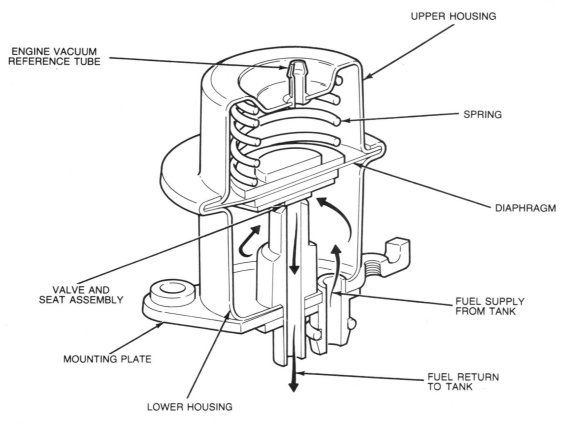

Fig. 10-10. The FPR used by Ford on the MPFI system uses vacuum and hydraulics to maintain pressure in the fuel system at a desirable level.

5. Now, look for a rubberized insulator ring on the back end of the filter. Remove it and lay it aside. Slide the filter out of the retainer, but as you do notice in which direction the arrow printed on the filter is facing. The arrow on the new filter has to point the same way. Lay the retainer aside and throw the old filter away.

6. Install the new filter by putting it inside the retainer with the arrow pointing correctly—that is, in the direction of fuel flow.

7. Install the rubberized insulating ring, but if the filter does not fit snugly in the retainer, get a new insulating ring.

8. Place the fuel filter and retainer in the bracket and tighten the retaining bolts 51 to 60 inch-pounds using a torque wrench.

9. Reattach the fuel lines, disconnect the fuel pressure gauge, and start the engine. Let it run for 30 seconds. Then check for gas leaks. If any are found, fix them.

THE FUEL PRESSURE REGULATOR

The Ford MPFI FPR (Fig. 10-10) is a hydromechanical device—that is, it combines hydraulics and vacuum (mechanics) to maintain a suitable pressure in the fuel system. Inside the metal housing of the FPR are two ports. They are a fuel supply port and a fuel return port. The fuel supply port is always open. Depending on how the engine is being operated, the fuel return port is partially or fully open.

Under idling or deceleration, a combination of intake manifold vacuum pulling up on a spring-loaded diaphragm inside the FPR chamber and fuel pressure pushing up on the diaphragm causes a seat over the fuel return port to rise and the port to open fully. This allows excess gas to bleed off and return to the fuel tank.

When vacuum in the intake manifold drops, as it would when the car is cruising along at a steady speed, only the pressure of gas being delivered by the fuel pump is present inside the FPR to raise the seat off the fuel return port. Intake manifold, being low, is not a fac-

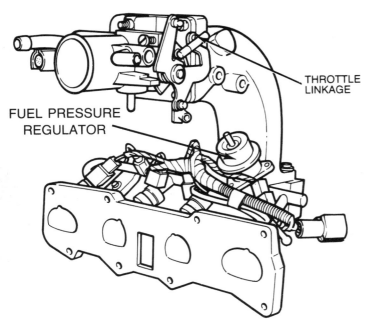

Fig. 10-11. This drawing points out the position of the FPR on the popular Ford 2.3-liter engine.

tor. Thus, the size of the port opening in the FPR is relatively small and a lesser amount of gas is allowed to return to the fuel tank.

If the FPR is damaged, the delicate pressure balance will be disrupted and the MPFI system will be thrown out of whack and cause an engine performance problem. This will happen if the diaphragm inside the FPR develops a crack or pinhole. If troubleshooting the system establishes that this has possibly happened (see pages 90–91), replace the FPR in this way:

1. Relieve fuel system pressure.

2. Disconnect the fuel delivery and fuel return lines from the FPR.

3. Undo the screws holding the FPR to the intake manifold (Figs. 10-11, 10-12). Discard the regulator plus the gasket and O ring that are on the underside of the regulator.

4. If the gasket or O ring has disintegrated so some material is sticking to the manifold, gently scrape that material off the manifold.

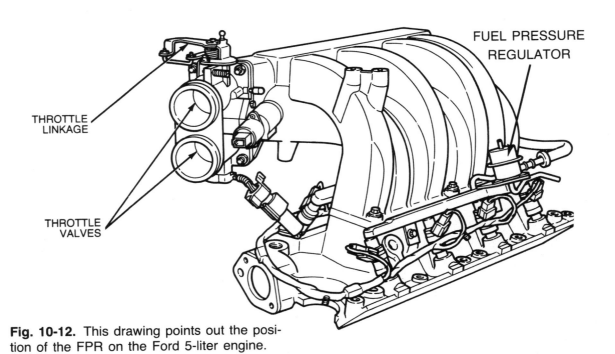

Fig. 10-12. This drawing points out the position of the FPR on the Ford 5-liter engine.

To make it easier, apply solvent to the material. Let the solvent soften the material; then gently scrape it off. Avoid hard scraping—you may damage the manifold.

5. Before installing the new FPR, spread lightweight oil on the O ring.

Caution: Do not use a silicone grease, which may work its way to the fuel injectors and clog them.

6. Place the new FPR on the manifold and tighten the retaining screws snugly. If you have a torque wrench that is calibrated in inch-pounds, so much the better. Tighten the screws 27 to 40 inch-pounds.

REPLACING THE FUEL INJECTORS

The steps outlined here refer to the Ford 2.3-liter 4-cylinder turbocharged engine equipped with MPFI. The engine you work on may not be as complicated; therefore, you might not have to go through all the steps noted here (Fig. 10-13).

Fig. 10-13. Depending upon how many other components are in the way, replacing the fuel injectors in a Ford engine could be an easy or difficult task. This injector is one that's easy to get at and therefore easy to replace.

1. Relieve fuel system pressure.

2. Disconnect the electrical connectors from the air bypass valve, throttle position sensor (TPS), injector wiring harness, knock sensor, fan temperature sensor, and coolant temperature sensor. As you do, mark each connector to identify it with the part to which it goes.

3. Disconnect intake manifold vacuum fitting connections by undoing the vacuum line fitting at the cast tube assembly, which is a tube through which air travels from the turbocharger to the throttle body, the rear vacuum line at the dash panel tree, the vacuum line to the exhaust gas recirculation (EGR) valve, and the vacuum line that goes to the FPR.

4. Disconnect the throttle linkage and the accelerator cable from the bracket. Tie the cable back so it's out of the way.

5. Remove the two bolts holding the cast tube assembly to the turbocharger; then remove the four nuts holding the air throttle body to the fuel charging assembly.

6. Separate the cast tube assembly from the turbocharger, making certain that you retrieve the gasket between the cast tube and turbocharger. Discard that gasket and get a new one.

7. Take the throttle body and cast tube from the car.

8. Now disconnect the positive crankcase ventilation (PCV) system by pulling the PCV hose off the fitting on the bottom of the upper intake manifold.

9. Drain the cooling system.

10. Loosen the hose clamp on the coolant bypass hose at the lower intake manifold and disconnect that hose.

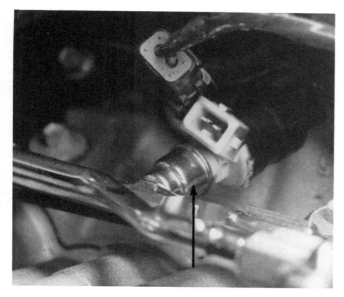

Fig. 10-14. To replace a fuel injector in a Ford engine once you reach it, disconnect the wire harness, undo the connector attaching the injector to the fuel rail (arrow), and rock the injector from side to side until it breaks away from the engine.

Fig. 10-15. The TPS is on the air intake (arrow).

11. Disconnect the EGR tube from the EGR valve by taking off the flange nut.

12. You can now proceed to work on the fuel injectors. Grasp a fuel injector connector and pull it off the injector (Fig. 10-14).

13. Grab the injector and rock it from side to side as you pull up on it. The injector will come out.

14. To install a new injector, notice that there are two O rings—one on top and one on the bottom. Spread a coat of lightweight oil on both. Do not use silicone grease. If you do, you stand a good chance of having that grease work its way inside the injector, which cause it to clog.

15. Using a twist and a push, seat the injector in the engine. Attach the connector.

After replacing all fuel injectors that have to be changed, put the fuel charging assembly together in reverse order to the way in which you took it apart.

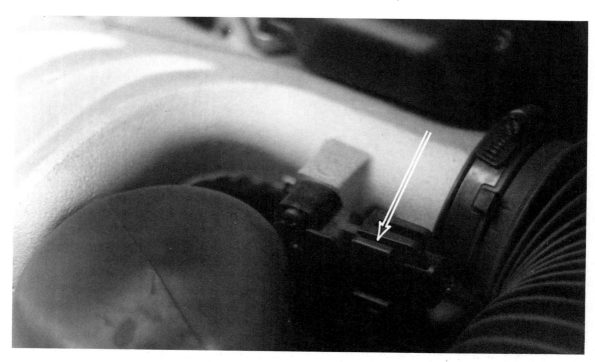

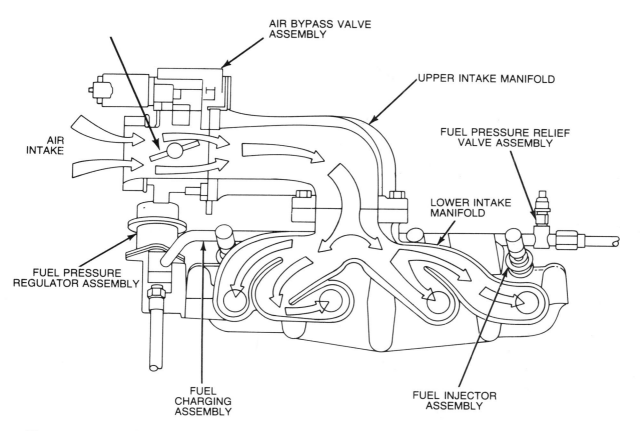

Fig. 10-16. The TPS monitors the position of the air intake valve (arrow).

THE THROTTLE POSITION SENSOR

The TPS is on the side of the air intake (Fig. 10-15). It is a potentiometer that monitors the position of the air intake valve inside the air intake and sends signals of varying intensity to the computer, telling it whether the air intake valve is partially open (normal operation), wide open (maximum acceleration), or closed (idling or decelerating) (Figs. 10-16, 10-17).

Fig. 10-17. Before concluding that the TPS is the cause of a problem, inspect the air filter. If it is clogged, the filter may be the reason for trouble.

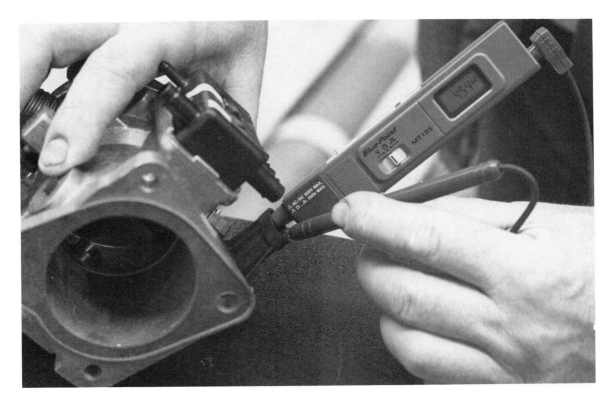

Fig. 10-18. Depending upon the engine, the TPS is rated at a specific ohm value. Before indiscriminately replacing the part, check it with an ohmmeter to see if it meets the value, which is specified in the service manual for the car.

The computer uses this information to get the fuel injectors to open and close at time intervals necessary for delivering the correct amount of gas for the way that the engine is operating—normal, maximum acceleration, idling, or deceleration (Fig. 10-18).

If troubleshooting makes you suspect the TPS as causing the performance problem, remove it by pulling the wiring harness from the TPS and undoing the two screws. When you install a new TPS, make sure the rotary tangs on the sensor fall into proper alignment with the throttle shaft blade, and that the sensor wires are facing down (Fig. 10-19).

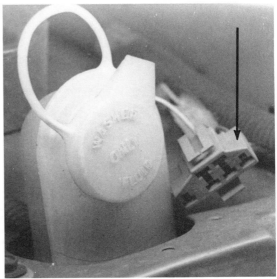

Fig. 10-19. All Ford engines with EFI have a diagnostic plug (arrow) in the engine compartment. It is for attaching a testing instrument called the STAR tester, which is a Rotunda tool available from Ford. The STAR tester lets you know at a glance if a problem you're experiencing is with the MPFI system or the engine computerized control system.

11

Chrysler Single-Point Electronic Fuel Injection System

Electronic Considerations

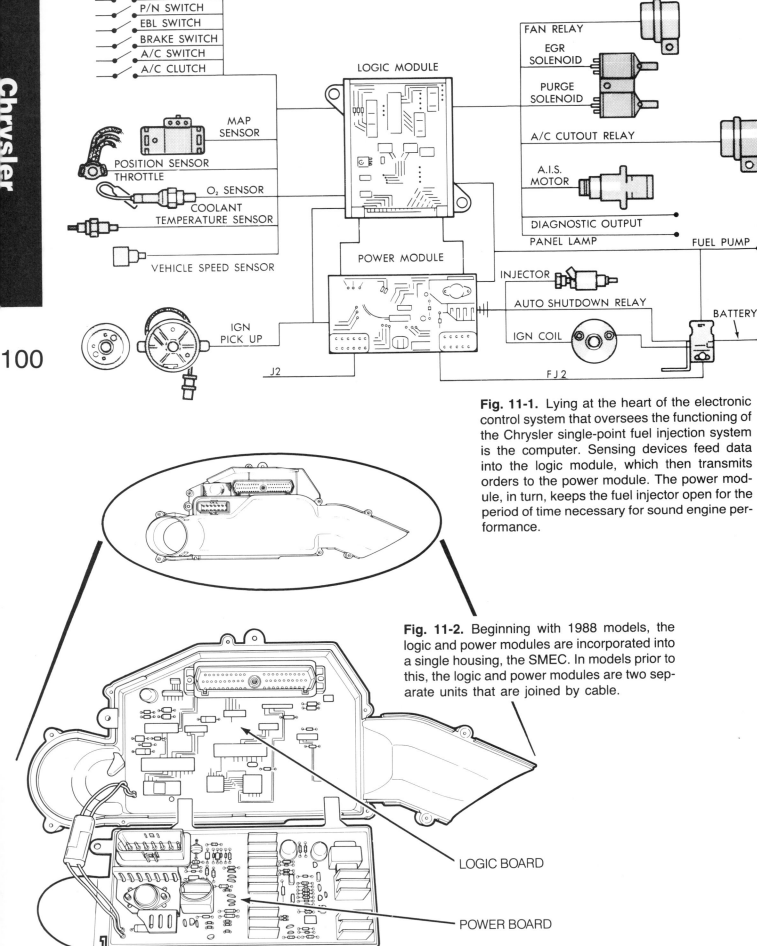

Fig. 11-1. Lying at the heart of the electronic control system that oversees the functioning of the Chrysler single-point fuel injection system is the computer. Sensing devices feed data into the logic module, which then transmits orders to the power module. The power module, in turn, keeps the fuel injector open for the period of time necessary for sound engine performance.

Fig. 11-2. Beginning with 1988 models, the logic and power modules are incorporated into a single housing, the SMEC. In models prior to this, the logic and power modules are two separate units that are joined by cable.

CHRYSLER USES A single-point (throttle body) fuel injection (TBI) system on all nonturbocharged 2.2- and 2.5-liter four-cylinder engines, and on some six- and eight-cylinder engines. The system is controlled by a computer (Fig. 11-1). To be 100 percent accurate, in pre-1988 models the system is controlled by two computers: One is a logic module that computes the needs of the engine based on various data it gets from the sensors; the other is a power module that carries out the commands of the logic module in order for the fuel injection system as well as the ignition and emissions control systems to work. Since 1988, both logic and power models are combined in a single computer called SMEC, for single module electronic control (Figs. 11-2, 11-3).

As with every modern-day EFI system, the Chrysler single-point system has one major purpose: to get engines to operate economically and reliably on a precise (14.7 to 1) air-to-gas mixture, so that the lowest possible exhaust emissions are given off.

THE COMPUTERS

In the discussion that follows, the dual computers in pre-1988 models are described. What they do and how they do it also apply to the integrated SMEC.

The logic module is a preprogrammed digital computer that regulates the fuel injection system, ignition timing, and emissions controls to meet conditions under which the vehicle is operating. To control these, the logic module receives data from a variety of sensors. Using this information, it computes how long the fuel injector should remain open to let gas spray into the engine and the moment when ignition will occur. It sends these commands to the power module, which carries out the orders.

Also, if there is a malfunction within the fuel, ignition, or emissions control systems, the logic module is able to "tell" a technician where the fault lies. Information regarding the fault is ascertained by checking a light on the car's instrument panel or by connecting a diagnostic readout instrument to a diagnostic plug.

Fig. 11-3. Here is what the SMEC looks like. If either module goes bad, the entire assembly has to be replaced.

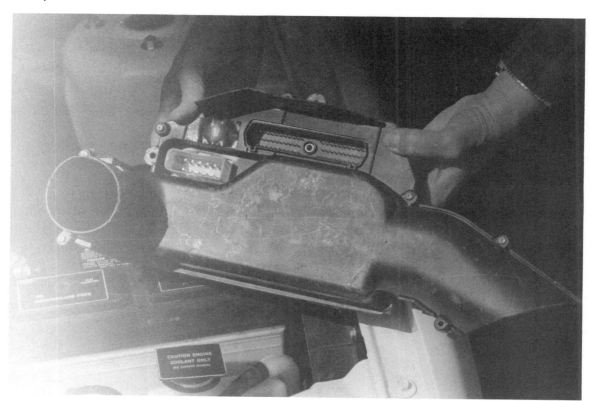

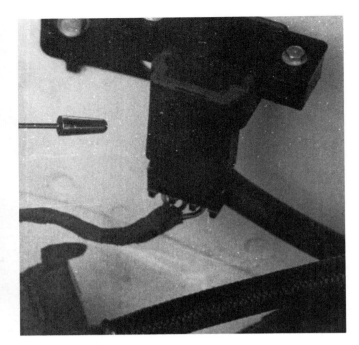

Fig. 11-4. The MAP sensor in a Chrysler vehicle with a single-point EFI system is usually mounted on a fender well.

The light or readout instrument displays a code that is cross-referenced with technical data issued by Chrysler to help a technician pinpoint the faulty circuit or component.

Along with carrying out logic module commands, the power module is involved with the safety of those riding in the vehicle. Contained in that module is an automatic shutdown (ASD) circuit. If the ASD circuit senses an imminent engine failure, it turns off the power module in less than one second. This action shuts down the fuel pump to prevent continued pumping of gas, which could cause the fuel system to load up, leak, and thus present a hazard.

THE SENSORS

Some sensors are found on all Chrysler models. Typical of these are the manifold absolute pressure (MAP) sensor, throttle position sensor (TPS), oxygen feedback (O_2) sensor, coolant temperature sensor (CTS), and speed sensor. Use the service manual for your vehicle to determine what others are installed in your car.

Keep in mind that even though components of the single-point EFI system in your vehicle are in good working condition, the engine will not perform properly if one of these sensors is faulty or a connecting wire is damaged. In fact, a problem is often blamed on the EFI when it really lies with a sensor. This can lead to unnecessary replacement of EFI parts in anticipation of correcting the problem.

Although this book is concerned with EFI, components such as these sensors—which are not directly engaged in the actual job of delivering gas to the engine—should be kept in mind. Being aware of them and what they do may help if a problem arises that appears to be EFI-related.

Manifold Absolute Pressure Sensor

The MAP sensor monitors manifold vacuum (Fig. 11-4). It is connected by a hose to a vacuum fitting on the intake manifold throttle body and by electric wire to the logic module. The sensor thus transmits data concerning manifold vacuum conditions to the logic module. This data is used along with other information received by the logic module to determine the proper air-to-gas ratio for driving conditions at the time.

Fig. 11-5. The TPS (pointer) in a Chrysler vehicle with a single-point EFI system is on the throttle body.

Throttle Position Sensor

The TPS is a resistor that emits varying amounts of electric current according to the positions the throttle valve assumes, and detects the rate of throttle position change (Fig. 11-5). The amount of voltage is transmitted to the logic module, which uses it along with voltage signals received from other sources to adjust the air-to-gas ratio to meet varying operating conditions and demands made during vehicle acceleration, deceleration, idling, and wide-open throttle operation.

Oxygen Sensor

The O_2 sensor, which is inserted into the exhaust manifold, monitors the oxygen content of exhaust gases. Based on the level of oxygen it detects, it produces a voltage that's sent to the logic module, which in turn signals the power module to modify the fuel injector "pulse" or duty cycle. The fuel injector is then commanded to open for a longer or shorter time depending upon operating conditions.

When a large percentage of oxygen is detected, which signifies that the fuel mixture is too rich, the O_2 sensor produces a low voltage and the injector is commanded to stay open for a shorter time. Conversely, when there is a lesser amount of oxygen present, indicating a leaner condition, the O_2 sensor produces a higher voltage to keep the injector open for a longer period.

Coolant Temperature Sensor

The CTS keeps tabs on engine operating temperature by monitoring the temperature of coolant (Fig. 11-6). This sensor, which is screwed into the thermostat housing, is connected to the logic module by electric wire.

Monitoring engine operating temperature is important for having an engine run well. When a cold engine is started, it requires a slightly richer-than-normal fuel mixture to produce a higher idling speed that will prevent stalling. As the engine warms up, the richer-than-normal fuel mixture, if it prevailed, could make the engine flood and stall. Therefore, the CTS "informs" the logic module to get the power

Fig. 11-6. The CTS in a Chrysler vehicle with a single-point EFI system is installed in a spot on the engine where it can come into contact with coolant. One such place is the thermostat housing.

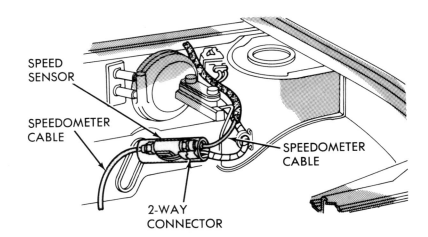

Fig. 11-7. The speed sensor in a Chrysler vehicle with a single-point EFI system is connected in series with the speedometer cable, so it can sense how fast the vehicle is traveling.

module to "ease off" the fuel injector and get it to operate to accommodate a warmed-up engine—which means stopping it from staying open for the length of time needed to support a cold engine.

Speed Sensor

This sensor is placed where it can sense the vehicle's rate of speed (Fig. 11-7). It is usually found at the spot where the speedometer cable is connected to the transmission.

The speed sensor is a microswitch that generates eight pulses for each revolution of the speedometer cable. The job of the speed sensor is to allow the logic module to differentiate between a closed-throttle deceleration condition and a normal closed-throttle idling condition when the vehicle is at a standstill. This allows the logic module to control the fuel injection system automatic idle speed (AIS) motor.

When deceleration takes place, the AIS motor is kept inactive by the logic module, because the speed sensor is informing the logic module that deceleration and not standstill idling is occurring. Conversely, at a standstill idle the speed sensor informs the logic module that the vehicle is not moving; therefore, the AIS motor should be activated to keep the engine running.

12 Chrysler Single-Point Electronic Fuel Injection System

Makeup of the System

Fig. 12-1. The throttle body (arrow) of a Chrysler single-point EFI system is mounted on the intake manifold in the same position a carburetor is mounted on an engine possessing a carburetor fuel system. As with a carburetor, a throttle body has a filter to trap impurities carried by air before air mixes with gas to form the fuel mixture.

THE HEART OF THE Chrysler single-point EFI system is a throttle body that's mounted on the intake manifold (Fig. 12-1). It is located in the same position where the carburetor is mounted on engines with carburetored fuel systems. The throttle body's functions are to meter the amount of air needed by the engine to operate, provide a means for air to mix with gas, atomize the mixture to make a combustible fuel, and provide the means for this fuel mixture to enter the engine (Fig. 12-2).

The key components of the Chrysler throttle body are the fuel injector, fuel pressure regulator (FPR), automatic idle speed (AIS) motor, and throttle position sensor (TPS) (Fig. 12-3). Remember, though, the throttle body is only one part of the Chrysler single-point fuel injection system. The system also includes elements that are involved in getting gas to the throttle body. These fuel delivery components are a fuel pump, fuel sock, and fuel delivery line. There is also a fuel return line. These components are discussed later in this chapter, but first let's describe the task of each of the parts attached to the throttle body.

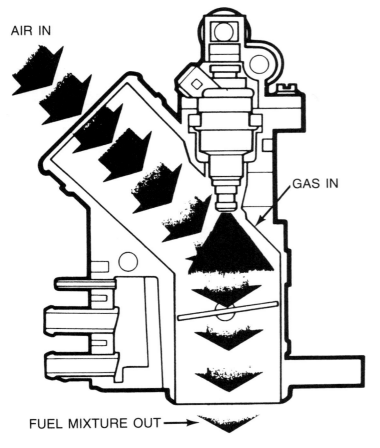

Fig. 12-2. As with throttle body systems used by GM and Ford, the throttle body of a Chrysler single-point EFI system mixes air and gas in the proper proportions. The mixture passes through the throttle valve into the engine.

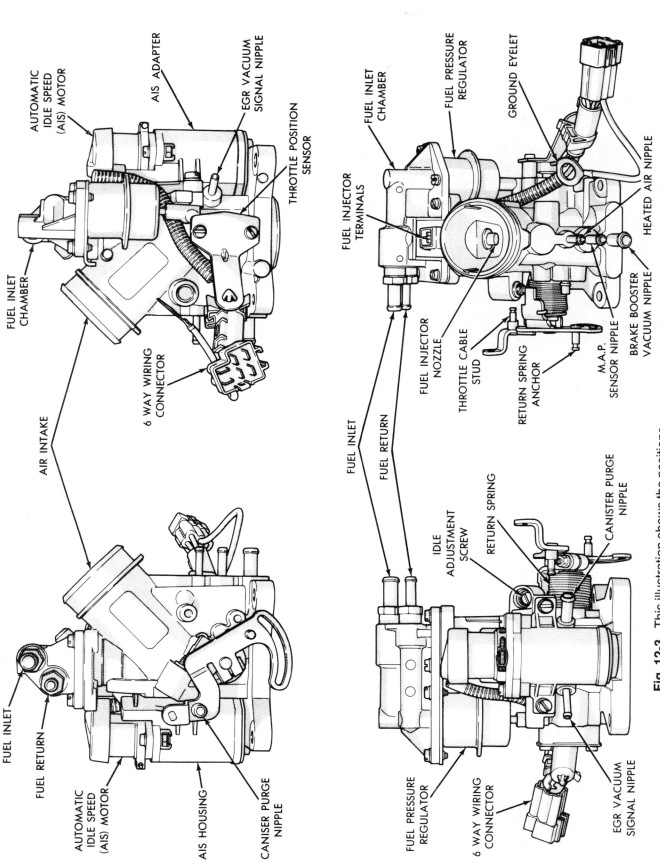

Fig. 12-3. This illustration shows the positions of the parts that make up the throttle body of the Chrysler single-point EFI system.

Fig. 12-4. The fuel injector in a Chrysler throttle body is protected by a cover (arrow).

THROTTLE BODY COMPONENTS

Fuel Injector

The Chrysler single-point EFI system used on 2.2- and 2.5-liter four-cylinder engines has one fuel injector. There are two fuel injectors used on 3.9-liter V-6, 5.2-liter V-8, and 5.9-liter V-8 engines that utilize the single-point system (Fig. 12-4).

The fuel injector is a precisely machined electric solenoid that opens and closes on command of the power module. However, the logic module governs how long the injector stays open and ejects gas, and how long it stays closed.

The injector gets power from the battery by way of the power module. Gas coming from the fuel delivery system first enters a fuel inlet chamber, which is attached to a fuel metering cover on top of the throttle body. The fuel metering cover also holds the FPR. Gas flows from the fuel inlet chamber into the fuel injector.

Fig. 12-5. The FPR (arrow) of a Chrysler single-point EFI system throttle body maintains pressure within the fuel system at a predetermined level. If the FPR malfunctions, the delicate balance will be disrupted and the engine will develop a performance problem.

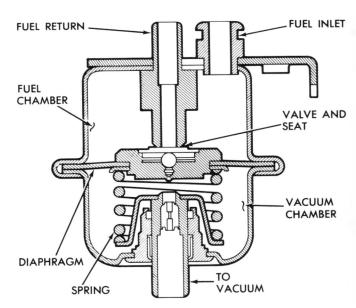

Fig. 12-6. This illustration of the inside of the FPR used by a Chrysler single-point EFI system plus the explanation in the text will help you to visualize how this important part works to maintain desirable pressure in the system.

When an electric pulse is applied to a wire inside the fuel injector, an armature in the base of the injector rises a short distance until it is stopped by a spring. Gas, which is under pressure, then passes through the injector to an outlet orifice (pintle) at the tip of the injector. The pintle is designed to allow a fine spray of gas in the shape of a hollow cone to spray from the injector and mix with air. The mixture enters the intake manifold and from there into the cylinders of the engine, where it is ignited.

Fuel Pressure Regulator

The FPR is a mechanical component positioned downstream of the fuel injector. Gas gets to the FPR after it has filled and satisfied the requirements of the fuel injector. The job of the FPR is to maintain pressure in the system at a constant 36 psi (Fig. 12-5).

If pressure at the fuel injector exceeds 36 psi, a diaphragm inside the FPR is pushed down to uncover a fuel return port (Fig. 12-6). Gas flows into this port and through a fuel return hose back to the fuel tank. By releasing fuel, pressure is relieved and controlled.

When pressure at the fuel injector falls below 36 psi, the tension exerted by a spring against the diaphragm causes the diaphragm to stay raised. The fuel return port stays closed and pressure returns to normal.

Automatic Idle Speed Motor

The AIS motor controls the flow of air into the throttle body. Data fed into the logic module from sensors are calculated by that module to allow the AIS motor to adjust the amount of air needed by the engine to run properly as it is idling. The logic module "instructs" the AIS motor to feed that amount of air to the throttle body, where it mixes with gas. The AIS motor adjusts the quantity of air according to need by increasing or decreasing the size of an air bypass opening on the rear of the throttle body.

FUEL DELIVERY COMPONENTS

As mentioned before, the system that delivers gas to the throttle body from the fuel tank consists of the fuel pump, which possesses a filter called a sock, a fuel delivery line, and an in-line filter between the fuel tank and fuel injector. These are instrumental in delivering gas from the fuel tank to the throttle body. Since the engine requires only a small portion of this supply, the remaining portion is returned to the fuel tank through a fuel return circuit, which consists of a fuel return line and fuel return check valve. The fuel return check valve, which is in the fuel tank, prevents gas from flowing back into the fuel return line.

You should be aware of these facts about the fuel delivery system of the Chrysler single-point EFI system:

1. All hoses employed in the system are made of special high-pressure material that meets Chrysler specification MS-EA-235. Do not use a hose made of any other material. Furthermore, install only fuel hose clamps that are specified by Chrysler for EFI usage—otherwise, there's a possibility that clamps will cut a hose. For these reasons it is best to purchase replacement fuel hoses and clamps from a Chrysler Motors dealer.

2. The in-tank fuel pump is a positive-displacement roller vane unit equipped with a permanent-magnet electric motor. It is part of an assembly that also includes the sock and a fuel damper, which reduces the level of noise made by the pump as it operates to keep noise from being audible to those in the car (Fig. 12-7). The sock is on the fuel pump inlet. It traps water and foreign

particles that get into the fuel tank before these contaminating agents can enter the pump and, consequently, the fuel delivery system.

The filter sock is self-cleaning and will seldom, if ever, plug. As water and particles are washed off the sock by gas that surrounds the fuel pump, they fall to the bottom of the fuel tank.

The fuel pump possesses two check valves. One valve, which is located on the inlet side of the pump, has the job of regulating the amount of pressure the pump can exert. This valve is designed to open and relieve pressure if the pump builds up an output of about 120 psi.

The second fuel pump check valve is located on the outlet side of the pump. Its job is to close and prevent any movement of gas out of the pump back into the fuel tank when the pump is not in operation. Gas is therefore immediately available to ensure quick engine starts. Current for the fuel pump is supplied from the battery through the ASD relay. See Chapter 11 for a description of the ASD.

Caution: The fuel pump is submerged in gas at all times. Therefore, it does not contain sufficient oxygen to form a combustible mixture as long as gas is in the tank. If you ever remove the pump for testing, be certain you don't apply voltage to it without first submerging it in a fluid. If you do, there's a possibility the pump may explode since gas or gas vapors may be trapped inside.

3. The in-line fuel filter is between the fuel tank and throttle body in the rear of the vehicle near the outlet end of the fuel tank. The in-line filter is made of a much finer material (paper, actually) than the coarse fuel pump sock. The in-line filter is a replaceable service part and should be changed when the engine displays a lean-fuel condition.

That is all there is to the Chrysler single-point EFI system, except for an air filter element that should be replaced every 52,000 miles unless the vehicle is operated under extremely dusty conditions. Then it is a good idea to inspect it every 15,000 miles to see if it is dirty. If it is, get a new one.

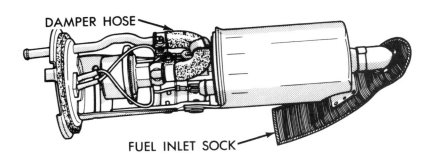

Fig. 12-7. As with all in-tank electric fuel pumps used to deliver gas to any type of EFI system, the fuel pump employed by a Chrysler EFI system is outfitted with a filter called a "sock" that is designed to trap particles carried by gas. This pump also uses a fuel damper to reduce noise the pump makes.

13

Chrysler Single-Point Electronic Fuel Injection System

Troubleshooting and Repair

According to Chrysler Motors, "Experience has shown that most complaints involving an electronic fuel injection (EFI) system can be traced to poor wiring and loose vacuum hose connections. A visual check will help spot these common faults and save unnecessary test and diagnosis time." Therefore, before troubleshooting the fuel injection system in a car that is demonstrating a performance problem, check wiring and vacuum hoses.

There are a number of areas where a bad electrical connection or a leaking or loose vacuum hose can exist in a Chrysler vehicle with EFI. Chapter 1 discusses procedures to use in tracking them down. Chrysler, however, recommends a particular order to follow:

1. Start by checking the various vacuum ports tapped into the throttle body. Look at the front face of the assembly and then at the rear. On the front of the throttle body, you will find a minimum of three vacuum ports. One holds the vacuum hose that connects to the MAP sensor, one holds the vacuum hose that supplies the brake booster, and the third holds the vacuum hose for the air cleaner heated-air door. Make sure hoses are securely connected to these vacuum ports, and that the hoses are in good condition over their entire lengths.

   On the rear of the throttle body you will find vacuum ports for the exhaust gas recirculation (EGR) valve and the purge valve of the fuel evaporation system charcoal canister. Again, check connections and condition of hoses (Fig. 13-1).

2. Proceed to the EGR and charcoal canister purge valve vacuum switches. In many cases one vacuum switch serves both the EGR and purge valves. It is a two-in-one switch with vacuum hoses for each component—EGR and purge valves—coming to it from vacuum ports on the throttle body. One side of the switch services the EGR valve; the other side provides vacuum for the purge valve. You can pinpoint this vacuum switch by tracing the vacuum hoses from the throttle body or by using the diagram that is included on the VECI decal. Make sure hoses are connected securely and are not leaking.

3. Check the hose and connections at the EGR valve (Fig. 13-2). Then turn your attention to the charcoal canister to make sure all of the hoses attached to the canister are secure and in sound condition (Fig. 13-3).

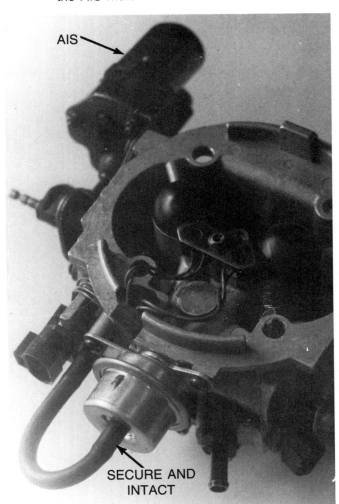

Fig. 13-1. It is essential that all vacuum hoses are securely attached on an engine equipped with a Chrysler single-point EFI system. If a hose fails or comes loose, the engine will falter. In this photograph, notice the position of the AIS motor.

Fig. 13-2. Be sure to test the EGR valve. Quite often the fuel injection system is blamed for a problem that's caused by the EGR valve.

Fig. 13-4. The PCV valve has been used to combat auto emissions since early in the 1960s. Don't overlook the possibility that an engine performance problem may be caused by a valve that has malfunctioned or by a PCV valve hose that has cracked or worked loose from a fitting.

4. Find the PCV valve. A hose extends from it to a vacuum port on the intake manifold. See that this hose is securely connected at each end and is not damaged (Fig. 13-4).

5. The air cleaner housing has several hoses attached to it. The large hose introduces heat into the housing from a heat chamber on the exhaust manifold; another is a vacuum delivery hose that extends from the housing to the throttle body; a third is an air duct hose extending from the housing to the power module. Check each hose carefully (Fig. 13-5).

Fig. 13-3. This illustrates what the charcoal canister of the fuel evaporation system looks like. It is buried deep in the engine compartment, often in the front near the fender housing on the right side.

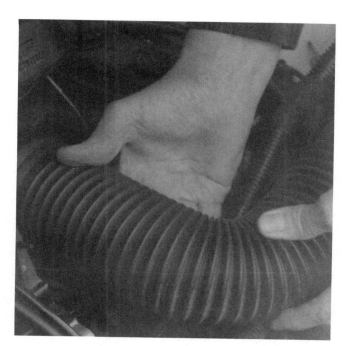

Fig. 13-5. A crack in one of the air hoses used on a Chrysler engine equipped with the single-point fuel injection system will cause the engine to falter. This is particularly tricky to track down since the crack is often hidden in one of the folds of an air duct.

6. That leaves wire harness connectors. Pull each connector apart and examine inside the two halves to verify that terminals aren't corroded or bent. If you find corrosion, clean it off with a small brush. If you spot bent terminals, straighten them with needle-nose pliers. Before pushing the halves firmly together, spread dielectric grease around the inside of each part if grease is present when you separate the halves. If grease is not present, leave the connectors dry. Finish the examination by inspecting wires for worn insulation. Worn insulation can be wrapped with electrician's tape.

The following are the connectors to inspect: the 21-way natural-color connector on the logic module; the 21-way black connector on the logic module; the 3-way connector at the MAP sensor; the 3-way and 1-way connectors at the ASD relay; the 12-way natural-color connector at the power module; the 10-way black connector at the power module; the 3-way connector at the EGR and purge solenoids; the 6-way connector at the AIS motor and TPS; the ground wire connection on the intake manifold; the 2-way connector at the fuel injector; the 1-way connector at the O_2 sensor; and the 3-way connector on the distributor (Fig. 13-6).

Fig. 13-6. Check electrical connectors. Pull each apart and look for corrosion and bent terminals. Don't overlook the connectors attached to the throttle body.

SELF-DIAGNOSIS

The logic modules of Chrysler vehicles equipped with single-point fuel injection systems are programmed to monitor various circuits associated with that system and will display codes to inform technicians of problems. Depending on the vehicle, the code is displayed by a flashing CHECK ENGINE light on the instrument panel or with a diagnostic readout instrument.

If the vehicle has a CHECK ENGINE light, it is activated by turning the ignition switch to the run position; then off; then run; then off; then run. This sequence puts the system into its self-diagnostic mode. This is followed by leaving the ignition switch on for 1 second; then off for 1 second; then on again.

The code is now displayed as two digits, such as 11, 23, or 34. If the light is displaying a code 23, it will flash twice quickly, then pause for a second, then flash three times quickly.

Trouble codes are a tip-off to an electrical or electronic problem that might otherwise be blamed on the EFI system. Therefore, when testing for a trouble code if none appears, concentrate on the EFI system. On the other hand, if you decide to investigate the EFI system first and find it operating as it should, look for a trouble code.

Since trouble codes change from year to year and differ from one model to another, the listing provided here may not be correct for your car. Consult the service manual for complete trouble code information. Here are several codes from the 24 possibilities that are programmed into the logic modules of 1984 models:

- **Code 11**—distributor circuit.

- **Code 13**—MAP sensor vacuum system.

- **Code 14**—MAP sensor electrical system.

- **Code 15**—vehicle speed sensor circuit.

- **Code 21**—O₂ sensor circuit.
- **Code 22**—CTS circuit.
- **Code 24**—TPS circuit.
- **Code 25**—AIS control circuit.
- **Code 31**—canister purge solenoid circuit.
- **Code 33**—air conditioning wide-open throttle cutout relay circuit.
- **Code 34**—EGR solenoid circuit.
- **Code 35**—fan relay circuit.
- **Code 41**—charging system.
- **Code 42**—ASD relay circuit.
- **Code 44, 52, 53, 54**—logic module.
- **Code 55**—end of message.

Fig. 13-7. The Chrysler-recommended method for releasing pressure in the fuel system of a single-point fuel injection system involves attaching two jumper wires to the terminals through which electricity is fed to the fuel injector (arrows).

TROUBLESHOOTING PROBLEMS IN THE EFI SYSTEM

To determine whether an engine performance problem is caused by an electrical or mechanical failure of a Chrysler single-point fuel injection system, make trouble light and fuel system pressure tests, respectively. You need an EFI-LITE and a fuel system pressure tester adaptable to a Chrysler single-point fuel injection system. For a suitable pressure tester, consult with an auto parts dealer. You may also get an EFI-LITE from that dealer or from the sources mentioned in Chapter 4, which also describes how to use the EFI-LITE.

Once you have determined that current is available to operate the fuel injector, turn your attention to the fuel system pressure test. Before making this test, however, be sure to release pressure in the fuel system to prevent gas from spraying when fuel line fittings are loosened. If you don't do this, gas can spray on you or on the surrounding area and create a hazardous situation.

The Chrysler single-point EFI system is under a constant pressure of approximately 36 psi. To release pressure in preparation for the fuel system pressure test, follow these steps:

1. Loosen the gas tank cap to relieve pressure in the tank.

2. Take off the air cleaner and disconnect the wiring harness connector at the fuel injector.

3. Notice that the fuel injector has two terminals (Fig. 13-7). Using a jumper wire equipped on both ends with alligator clips, attach one end of the jumper to one of the terminals—it doesn't matter which one. Attach the alligator clip on the other end of the jumper to a nearby bolt or bracket to ground the injector.

4. Now, connect a second jumper wire—a fairly long one—to the free terminal of the

fuel injector. Bring the other end of this jumper to the car's battery and hold it against the positive post of the battery for exactly 10 seconds. Gas will spray from the fuel system via the injector, which will relieve fuel system pressure.

5. Remove both jumper wires and reconnect the wire to the injector. No pressure should remain in the fuel system, but be careful anyway as you disconnect fuel system lines to do the fuel system pressure test. Wrap a rag around the connector as you loosen it.

Note: This procedure is the one recommended by Chrysler. Another way to release pressure is to pull apart the fuel pump electrical connector on the outside of the gas tank. Start the engine. When it stalls, crank it for a few seconds more. Reconnect the two parts of the fuel pump electrical connector.

Caution: Every time you allow fuel system pressure to build up as you perform the various steps involved in doing a thorough fuel system pressure test and other services, you must release pressure again before disconnecting another part of the fuel system.

MAKING THE FUEL SYSTEM PRESSURE TEST

To do the fuel system pressure test, remove the metal fuel delivery line from the fuel intake hose attached to the throttle body (Figs. 13-8, 13-9). Connect the fuel system pressure tester between this hose and the fuel delivery line (Fig. 13-10). Start the engine and record the reading shown on the pressure tester. That's all there is to it, except for interpreting readings.

A reading of 36 psi is normal. If you get this or within a couple of psi of this, remove the fuel system pressure tester and reattach the fuel intake hose to the fuel delivery line.

Start the engine. Gas should spray into the throttle body in the form of a well-defined cone pattern. As you watch the injector, have someone turn off the engine. Not even one drop of gas should leak from the injector. If it does, replace the injector. If the fuel injection system meets these tests, it is operating perfectly. The performance problem you are experiencing is with another part of the engine.

(Incidentally, if you use the metric scale of the fuel system pressure tester, a normal reading is 260 kPa, give or take 14.)

Suppose the fuel system pressure tester reading you get is more than a couple of psi below the normal 36 psi. In this case, locate

Fig. 13-8. There are two fuel hoses connected to the throttle body. The one shown here is the fuel intake hose. It has the larger diameter. Work with this hose when making the fuel system pressure test.

the in-line fuel filter. Disconnect the fuel line coming into the fuel filter from the fuel tank and connect the fuel pressure tester between the in-line fuel filter and the end of the line coming from the fuel tank. Start the engine and note the reading. If the fuel pressure reading now records normal, there is a restriction inside the fuel filter. You can correct the engine performance problem by installing a new fuel filter.

Suppose there is no change in the fuel system pressure reading; that is, it stays below 36 psi. Then find the FPR and identify the hose that extends from it to the fuel tank. This is the fuel return hose. Squeeze the hose gently and record the reading given by the fuel pressure tester. If pressure rises, the malfunctioning part you are looking for is the FPR, and you should install a new one. If there is no change in pressure, the trouble is inside the fuel tank with a clogged sock or a defective fuel pump. Have the fuel tank removed from the car to test the pump and make sure the sock is clean.

You may get another fuel system pressure test result when the fuel pressure tester is connected between the fuel intake hose and the throttle body—a pressure that is above the normal 36 psi. If this happens, disconnect the fuel return hose from the FPR, connect a substitute hose to the FPR, and aim the end of the substitute hose into a can. Start the engine. If the pressure reading does not fall to normal, the FPR is faulty. But if pressure does fall to normal, the fuel return hose is obstructed.

Fig. 13-9. The other hose connected to the throttle body is the fuel return hose. It is through this hose that excess gas flows back to the gas tank.

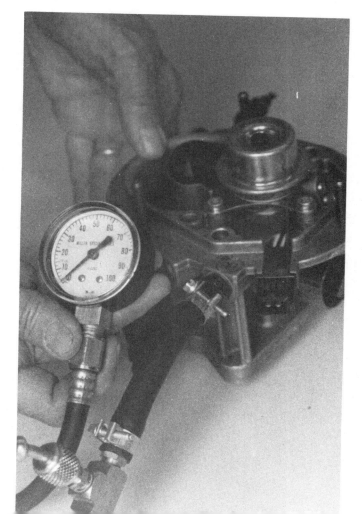

Fig. 13-10. Connect the fuel system pressure tester between the fuel intake hose and the fuel delivery line.

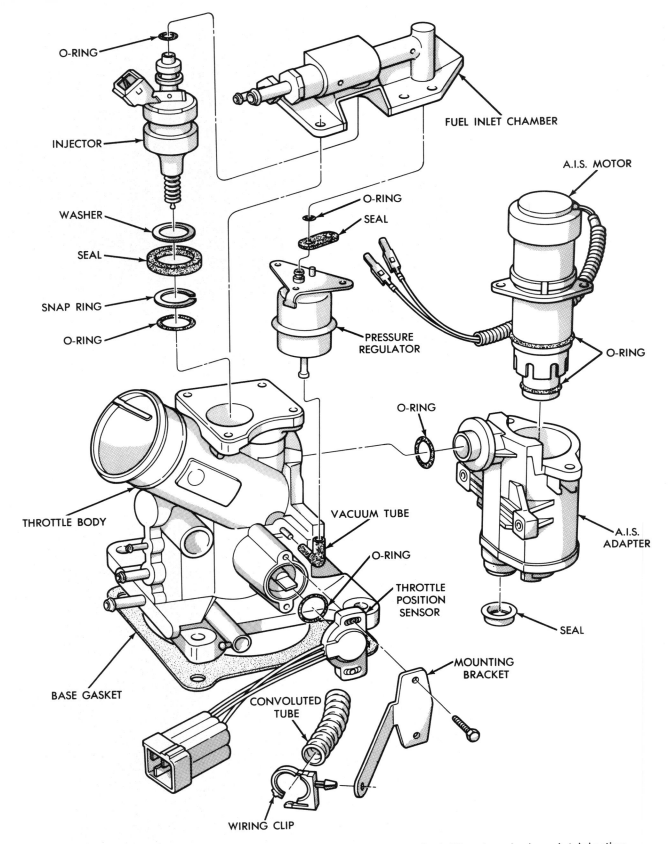

Fig. 13-11. This illustration showing the breakdown of a typical Chrysler single-point injection system throttle body into its various components will provide guidance when you have to replace a part. Notice the various seals and O rings. Be sure to use new O rings and seals when installing a new part.

REPLACING TBI PARTS

The descriptions that follow regarding replacing the in-line fuel filter, fuel injector, and FPR may not be 100 percent accurate for your particular Chrysler model. The methods of installing these parts vary from one model year to the next, but the data is correct in a general sense. It is provided to demonstrate that replacing one of these parts is not difficult (Figs. 13-11, 13-12). You do not have to remove the throttle body from the car to do it.

1. Let's begin with the in-line fuel filter: Disconnect the battery negative cable and relieve fuel system pressure. Loosen the clamps holding fuel hoses to the filter, wrap a towel around the hoses, and pull off the hoses. Discard the clamps. Loosen the screw holding the collar around the filter and slide the filter from the collar. Slide a new filter into place, secure it to the collar, slide new clamps onto the hoses, push the hoses firmly onto the filter, and tighten the clamps. Reconnect the battery negative cable, start the engine, let it run for a minute, turn it off, and check around hoses for leaks. If there is a leak, tighten the clamp a little more, but do not overtighten. You may distort the filter nozzle.

2. To replace the fuel injector in one type of Chrysler throttle body, disconnect the battery negative cable and release fuel system pressure. Remove the screws that hold the fuel inlet chamber to the throttle body. These are probably Torx screws, which look like Allen-head screws. You should use a Torx driver. If you use an Allen wrench, you will damage the screws.

   At the base of the FPR you will find a short piece of vacuum tube that extends to the throttle body. Disconnect it from the FPR. Wrap a towel around the fuel inlet chamber and lift the chamber holding the injector off the throttle body. Now pull the injector out of the chamber. To install a new fuel injector, just reverse these steps, but be sure to coat O rings with petroleum jelly

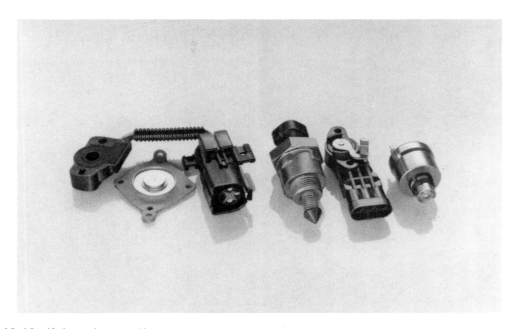

Fig. 13-12. If there is a malfunction with the throttle body, you do not have to replace the entire unit. Once the faulty part is identified, replace only that one. Parts are available from Chrysler dealers and from independent auto parts dealers.

and seat the injector firmly into the fuel inlet chamber. Check for fuel leaks after the job is completed.

To replace the fuel injector in another type of Chrysler throttle body, disconnect the battery negative cable and release fuel system pressure. Then, proceed as shown in Figs. 13-13 to 13-17.

3. If the fuel system pressure test reveals a faulty FPR, disconnect the battery negative cable, release fuel system pressure, and undo the fuel inlet and fuel return lines from the regulator. Undo the Torx screws that hold the FPR to the fuel inlet chamber, pull off the vacuum tube that extends between the FPR and throttle body, and remove the faulty FPR from the throttle body (Fig. 13-18).

To install a new FPR, press the part into its housing in the throttle body. Install the three Torx screws and tighten them securely without overtightening. Finally, re-connect the vacuum tube and fuel inlet and fuel return lines, connect the battery, start the engine, let it run for a minute or two, and check to see that there are no fuel leaks.

TROUBLESHOOTING THE AUTOMATIC IDLE SPEED MOTOR

The AIS motor on the throttle body is controlled by the logic module. It adjusts the air portion of the air-to-gas mixture that flows through the throttle body into the engine according to engine load and ambient temperature conditions. A faulty AIS will cause an erratic idling speed, an idle speed that is too fast, or an idle speed that is too slow.

If engine performance takes a turn for the worse, you can quickly determine if the AIS is working properly by letting the engine run at idle and having an assistant behind the wheel alternately turn the car's air conditioner on and off. Place a finger on the AIS. As the air con-

Fig. 13-13. Undo the screw that holds the protective cap over the fuel injector.

Fig. 13-14. Remove the cap. As you do this, the electric female terminals in the cap and the injector male terminals are separated.

Fig. 13-15. If you can't grab and pull the fuel injector out by hand, place a screwdriver on each side of the injector to pry it from its seat.

Fig. 13-17. Align the female terminals in the cap with the male terminals of the injector. Then press the cap into place.

Fig. 13-16. Before installing a new injector, make sure the O ring that fits around the bottom end of the injector is in place.

Fig. 13-18. The FPR is held to the throttle body by screws. To replace this important part, remove the screws.

ditioner goes on and off, you should feel the motor inside cycling on and off to provide a constant idling speed. If the part is not cycling, replace it as follows:

1. Turn off the engine and disconnect the battery negative cable.

2. Pull apart the two electrical connectors that attach the AIS to the 6-way connector of the throttle body. Note how both connectors are attached, so you can attach the connectors of the new AIS in the same way.

3. Remove the two screws holding the AIS to its housing and remove it from the housing.

4. Install a new AIS, making sure it possesses the two O rings necessary to seal it properly in the housing.

Note: Coat O rings with petroleum jelly.

14 Chrysler Multiport Fuel Injection System

An Overview

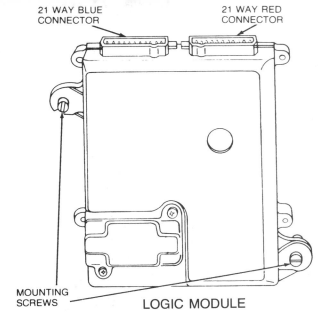

LOGIC MODULE

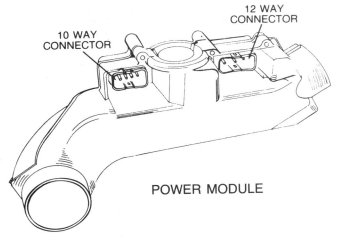

POWER MODULE

Fig. 14-1. Engines in cars manufactured prior to 1987 use separate electronic modules called the logic and power modules to control the functioning of the Chrysler MPFI system. The two are connected by cables.

Fig. 14-2. This device (arrow) inside the distributor of a Chrysler engine using MPFI measures engine speed and transmits the data to the logic module.

A MULTIPORT FUEL INJECTION (MPFI) SYSTEM, which is characterized by one fuel injector for each cylinder an engine possesses, was used on a limited number of vehicles bearing the Chrysler nameplate before 1987. For the most part, these were engines equipped with turbochargers. Beginning with the 1987 models, the application was expanded. In addition to four-cylinder engines with turbochargers, most naturally aspirated (that is, nonturbo) six- and eight- cylinder engines used by Chrysler have been equipped with MPFI.

THE ELECTRONICS

In pre-1987 models, the control of the MPFI system is maintained by two electronic modules called the logic and power modules (Fig. 14-1). These are separate units and are the same as those used to control the Chrysler single-point fuel injection system. The functions of the logic and power modules are described in Chapter 11.

In 1987 and newer Chrysler models, the control of the MPFI system is maintained by the single module engine controller (SMEC). This unit—call it a computer if you wish—incorporates the logic and power modules into one assembly. The function of each module is the same as in vehicles that have separate logic and power units.

Chrysler describes its MPFI system as a speed–density fuel injection system. Those two words—speed and density—refer to engine speed (in rpm) and the density of the air an engine ingests. Both are important for proper functioning of the MPFI system and therefore of the engine.

Engine speed is measured by an optical distributor that uses reference and synchronization pickups (Fig. 14-2). The two pickups—reference being the one that actually monitors engine speed, and synchronization the one that calibrates spark timing vis-à-vis speed—ensure delivery of ignition sparks to each cylinder at the precise time necessary to ignite the fuel mixture. Air density is measured

Fig. 14-3. This is the MAP sensor, which keeps tabs on air density. Data it accumulates are transmitted to the logic module.

by the manifold absolute pressure (MAP) sensor (Fig. 14-3).

Speed and density are monitored by the logic module, which then controls the time (pulse width) the fuel injectors remain open to let gas spray into the engine. At the intake valves gas mixes with air, and this properly proportioned fuel mixture then flows into the cylinders, where it ignites and burns.

There are three other components that are important to proper engine functioning: a throttle position sensor (TPS) (Fig. 14-4), coolant temperature sensor (CTS) (Fig. 14-5), and the charge temperature sensor (Fig. 14-6). Interestingly, these and all of the other sensors except one can fail and the engine

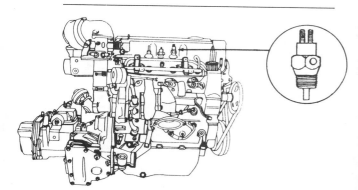

Fig. 14-6. The charge temperature sensor gathers data concerning the temperature of the air being taken into the engine through the air plenum. Data are transmitted to the logic module.

Fig. 14-4. The TPS is attached to the throttle body and monitors the position of the throttle valve. Data it gathers are transmitted to the logic module.

Fig. 14-5. The CTS is screwed into the engine where it can come into contact with coolant. Data it gathers are transmitted to the logic module.

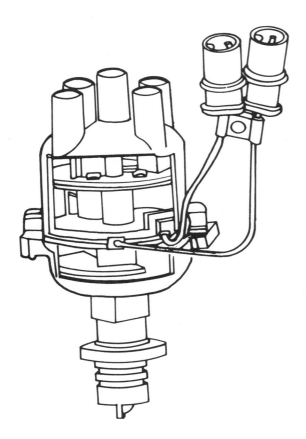

Fig. 14-7. If the engine runs badly, one or more of the sensors shown in Figs. 14-4, 14-5, and 14-6 may be bad. However, if the engine doesn't run at all, the problem might be a faulty rpm-sensing pickup located inside the distributor.

will still continue to run, although badly, since it will be in the so-called limp-in mode. The limp-in mode allows the engine to run and the vehicle to keep going until you reach help. The part of the electronic system that will cause the engine not to run if it fails is the rpm-sensing (reference) pickup in the distributor (Figs. 14-7, 14-8).

If your Chrysler engine develops a performance problem, first troubleshoot the vacuum system and electrical connecting points (Chapter 1), and then the fuel injection system (Chapter 15). If the cause of the trouble is not uncovered, the next step is to delve into the electronics (Fig. 14-9). For guidance in doing this, refer to the service manual for your specific vehicle.

BREAKING DOWN FUEL PRESSURE

The Chrysler MPFI system relies on an electric fuel pump located in the gas tank (Fig. 14-10). The pump that's used in most vehicles has the capability of operating at a maximum of 100 psi and delivering 30 gallons of gas per hour.

Fig. 14-8. Use an ohmmeter to determine if the rpm-sensing pickup inside the distributor is bad. If the test shows a resistance in excess of that given in the car's service manual, replace the pickup.

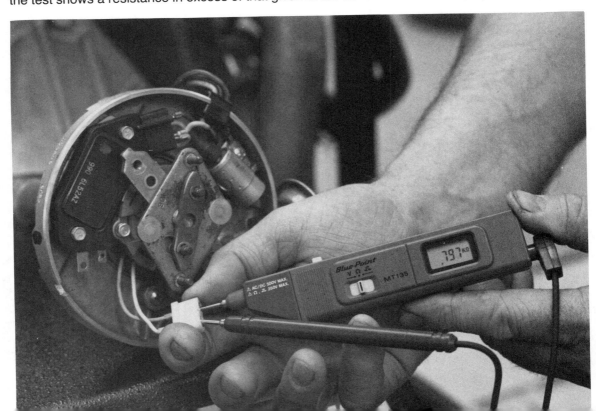

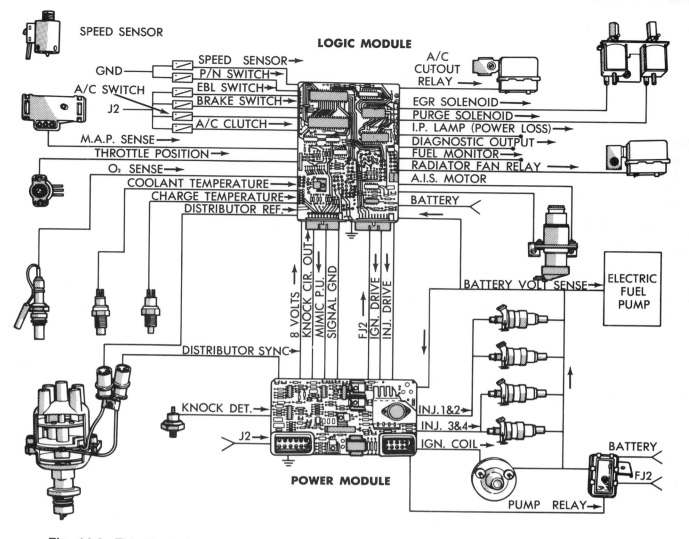

Fig. 14-9. This illustration provides an overview of Chrysler's electronic control system, which governs the length of time fuel injectors remain open to feed gas to the engine.

However, certain parts are incorporated in the MPFI system to reduce the maximum pressure capabilities of the pump—100 psi is much too much.

One of these devices is a pressure relief valve. Another is the fuel pressure regulator (FPR).

The pressure relief valve is incorporated within the fuel pump housing. It opens to release gas back into the fuel tank and thereby reduce maximum fuel pump pressure from 100 to 80 psi. The FPR does the rest. It maintains pressure in the system at whatever specification it is set to maintain: That could be 53, 48, or 30 psi (see below).

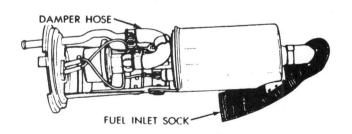

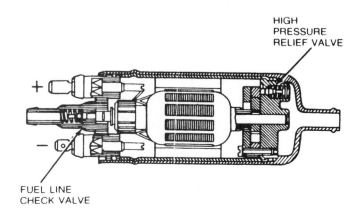

Fig. 14-10. The in-tank electric fuel pump used by the Chrysler MPFI system can deliver gas at a rate up to 30 gallons per hour. However, a high-pressure relief valve, check valve, and the FPR keep delivery at a level that coincides with the needs of the engine.

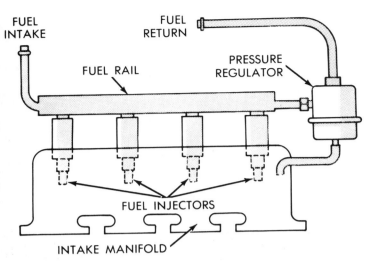

Fig. 14-11. The FPR on a Chrysler engine without a turbocharger is located on the fuel rail. Its purpose is to maintain a consistent pressure within the fuel system.

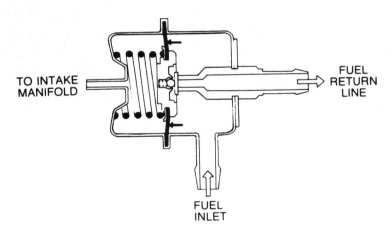

Fig. 14-12. When the predetermined pressure in the fuel system is in danger of being exceeded, a diaphragm in the FPR opens a port leading to the fuel return line so excess gas can flow back to the gas tank.

With MPFI systems used on four-cylinder engines having turbochargers, the FPR is positioned on the throttle body. On other engines with MPFI systems, the FPR is on the fuel rail (Fig. 14-11). The fuel rail is the elliptical line through which gas is delivered to the fuel injectors.

Note: The term "throttle body" as it applies to the MPFI system is not used in the same context as when it is used in referring to the Chrysler single-point EFI system (Chapters 11 to 13). The throttle body of the MPFI system is the air intake for the system. It monitors air density. Attached to the throttle body are the TPS and automatic idle speed (AIS) motor.

The FPR is a mechanical component that maintains a constant pressure at the fuel injectors. It does this by means of a spring-loaded diaphragm that controls access to a fuel return port. Connected to this port is a fuel line that returns gas not needed at the time to the fuel tank. It works this way:

The fuel pump delivers to the fuel injectors an amount of gas that causes injectors to overflow, sending excess into the FPR. As long as the pressure exerted by this gas is below the level at which the FPR is set—30 psi, say—the spring-loaded rubber diaphragm stays in place over the fuel return port and no gas can flow back to the fuel tank through the fuel return line. However, when the gas pressure builds up in the regulator to that setting, it lifts the spring-loaded diaphragm off the fuel return port (Fig. 14-12). Gas flows through the port back to the gas tank until the pressure is relieved and the spring-loaded rubber diaphragm drops to close the fuel return port. The spring-loaded rubber diaphragm is in constant motion, moving alternately to open and close the port to keep the pressure within the fuel system more or less constant. (*Note:* Pressure may vary from specification by a couple of psi.)

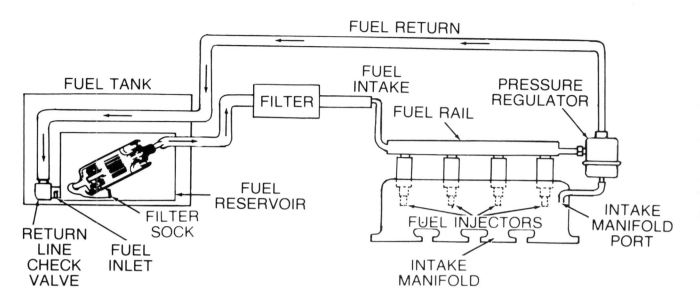

Fig. 14-13. Before gas gets to the fuel injectors, it is filtered by a sock in the gas tank and an externally mounted in-line fuel filter. This illustration will provide guidance to identify the various parts of the Chrysler EFI system.

As pointed out, the amount of pressure that is maintained in the fuel system by the fuel FPR differs from one engine model to the next. With 2.2-liter engines equipped with turbochargers, specified fuel pressure in the MPFI system is set at 53 psi. The fuel pressure for pre-1988 3.0-liter engines equipped with MPFI is 30 psi. In 1988 models with 3.0-liter engines, the setting is 48 psi. To determine the exact pressure the MPFI system in your vehicle is supposed to maintain, consult the service manual.

ROLE OF THE FILTERS

On its way from the fuel tank to the fuel injectors and FPR, gas passes through two filters that trap any contaminating agents. One filter is called a sock. It is in the fuel tank and is part of the fuel pump assembly. The other filter lies outside the fuel tank and is integrated into the fuel delivery line between the fuel tank and fuel injectors. This filter is called an in-line fuel filter (Fig. 14-13).

The purpose of the sock is to trap large particles that could be drawn into and damage the fuel pump. It is a coarse mesh filtering element that will not trap tiny particles. Furthermore, even if it traps a substantial number of large particles, the sock seldom clogs, because it is self-cleaning. As gas washes over the sock, any particles that settle on it are washed off and drop to the floor of the gas tank.

The one time a sock can get in trouble is if the vehicle owner is in the habit of letting the fuel tank run low on gas. Doing so curtails the washing effect of a full or nearly full tank. If this happens often enough, the large particles that are usually washed off the sock will cling to it. Then, the flow of gas is hampered and the engine develops a drivability problem.

Caution: In a vehicle possessing a fuel injection system, you shouldn't allow the gas level in the tank to drop below one-quarter full.

The purpose of the in-line fuel filter is to trap particles carried by the gas that escape the sock. The in-line filter is easily replaced if it gets plugged, a condition characterized by the engine demonstrating a lean-fuel symptom, such as hesitation, stumbling, or stalling.

Before indiscriminately replacing the in-line fuel filter, however—one can cost $30—conduct a fuel pressure test to establish whether there is a drop in pressure within the fuel delivery system (Chapter 15). If there is, it is reasonable to assume that the in-line fuel filter is clogged.

ROLE OF THE FUEL INJECTORS

In addition to the fuel pump, sock, in-line fuel filter, and FPR, the remaining parts of the MPFI system that deliver gas to the engine are the fuel injectors. There are four injectors in a four-cylinder engine, six if the engine has six cylinders, and eight if the engine is a V-8.

The fuel injectors are electrically operated solenoid valves that get power to function from the power module. However, they are controlled by the logic module, which establishes how long injectors should stay open. The logic module makes this determination based on data it receives from the various sensors. The logic module then instructs the power module to energize the injectors for whatever period of time is necessary to ensure sound engine performance.

When the power module delivers electric current to the injectors, an armature and pintle assembly in the tip of each injector moves up against a spring. This action opens an orifice in the tip through which gas sprays into the engine, where it mixes with air coming into the engine through the throttle body (Fig. 14-14).

In many Chrysler engines with MPFI, fuel injectors work one at a time. The injector setup in the 3.0-liter six-cylinder engine, however, is different. The six injectors work in pairs with each pair energized every 720 degrees of crankshaft rotation. The fuel injectors servicing cylinders 4 and 5 work as a team, those servicing cylinders 2 and 3 work together, and those for cylinders 1 and 6 operate simultaneously. Therefore, each cylinder gets one squirt of gas every four cycles. This kind of MPFI setup, which other automakers also use, is often called sequential fuel injection (SFI).

There is one important characteristic to keep in mind relative to troubleshooting the Chrysler multiport SFI system: If *one* injector of a pair stops working, the engine *may* shut down. If a *pair* of injectors is lost, the engine *will* shut down. One way of knowing if a single injector is causing engine die-out is this: After the engine dies it can be restarted, but stalls within 30 seconds as the car is traveling. If a pair of injectors is lost, the engine can't be restarted.

GETTING AIR

Finally, let's discuss the air intake system of the MPFI system (Figs. 14-15, 14-16). A Chrysler engine with MPFI has a throttle body

Fig. 14-14. The fuel injectors are electrically operated. When current is received, the pintle is raised off its seat and gas is sprayed into the engine.

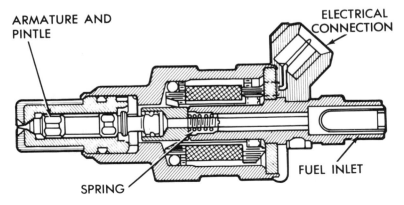

Fig. 14-15. Air gets to the cylinders of a Chrysler MPFI system through this housing, which is called the intake manifold or air plenum. Air travels through a filter housing, duct, and the throttle body before entering the air plenum.

through which it takes in air. The throttle body possesses an AIS motor and a TPS.

As the name implies, the function of the AIS is to ensure that the engine idles properly both when it's cold and when it's hot. The AIS, which is controlled by the logic module, adjusts the amount of air entering the engine by opening or closing an air bypass in the throttle body. The AIS is not a factor in troubleshooting unless the engine problem involves stalling at idle speed or when the engine is decelerating with your foot off the accelerator pedal.

The TPS is an electrically operated resistor attached to the throttle body. It senses the angle at which the throttle valve inside the throttle body is positioned. The throttle valve is a blade through which air passes into the engine as the driver presses the accelerator pedal.

The TPS creates a voltage, the strength of which depends on the angle of the throttle valve. This voltage is transmitted to the logic module. In conjunction with data transmitted by other sensors, this information is used by the logic module to adjust the amount of air that mixes with the gas to meet the varying conditions under which an engine must operate—specifically, acceleration, deceleration, idling, cruising, and wide-open throttle operation.

Chapter 15 that follows explains procedures involved in troubleshooting components of the Chrysler MPFI system to determine whether any of them is responsible for a performance problem.

Fig. 14-16. The throttle body houses the throttle valve. The angle of the throttle valve is established by the driver's foot on the accelerator pedal. As the accelerator pedal is pressed, the throttle cable moves. In so doing, it sets the position of the throttle valve.

15

Chrysler Multiport Fuel Injection System

Troubleshooting and Repair

Fig. 15-1. In checking that the electrical units in a car with an MPFI system are functioning properly, don't overlook the key electrical connection of all. Make sure battery cables are clean and secure.

ALTHOUGH TROUBLESHOOTING THE Chrysler MPFI system is similar to troubleshooting the GM and Ford MPFI systems, there are some differences. In fact, there are slight variations in servicing the Chrysler MPFI system because of changes incorporated into vehicles from one model year to the next. However, the information in this chapter will help you over the hurdles.

The MPFI system will malfunction if there is an electrical or mechanical failure. Your job is to determine if one has occurred and test the particular components involved to uncover the culprit—it is not a difficult task.

Fig. 15-2. The FPR (arrow) is connected to the intake manifold by a hose. It uses vacuum created in the manifold to maintain a consistent pressure in the fuel system. Make sure the vacuum hose is in sound condition and secured. If it isn't, pressure will be disrupted, and the engine will act up.

THE ELECTRICAL STORY

Chrysler contends that many complaints regarding a malfunction with the MPFI system are caused by an electrical failure. The best tools you have available for uncovering an electrical problem are your eyes, but before we describe what to look for, let's ask this question: How do you find out if the cause of the problem is electrical?

In Chapter 4 we discussed finding an electrical fault in the GM EFI system by using the EFI-LITE. This easy-to-use device can also reveal if you have an electrical problem with the Chrysler MPFI system.

Before ordering the tool, disconnect an electrical connector from one of the fuel injectors. Make a sketch of the terminals of the injector and note whether they are blades or prongs. You need this information along with the make, year, and engine nomenclature relative to the vehicle to be sure of getting the correct EFI-LITE. If you cannot obtain this instrument from an auto parts dealer, refer to Chapter 4, where companies selling the EFI-LITE are listed.

To use the EFI-LITE, disconnect each connector from its fuel injector, plug the EFI-LITE into the connector, turn on the ignition switch, and watch the EFI-LITE. It should emit pulsating flashes. If there is no light or if the light glows steadily, you have an electrical problem. However, if the light pulsates, forget "electric" and start thinking "mechanical."

THE ALL-IMPORTANT VISUAL INSPECTION

Suppose the EFI-LITE reveals an electrical malfunction. In hopes of finding it, inspect each electrical connector you can find for damage (Fig. 15-1).

Although electrical connectors and connections will vary between the Chrysler four-cylinder engine with MPFI, which is described here, and six- and eight-cylinder Chrysler engines with MPFI, there are many connections which are common between engines. The outline of electrical connectors for the four-cylinder engine provided here will give you an idea of where to look. In addition to these, don't forget to make sure vacuum hose connections are tightly secured, especially at the throttle valve, PCV valve, turbocharger, FPR (Fig. 15-2), EGR valve, and MAP sensor (Chapter 1).

The following are the electrical connectors that should be examined:

■ The six-way throttle body connector.

■ The ground connection for the main electrical harness going to the fuel injectors. On the four-cylinder engine with turbocharger and MPFI, this ground is in the form of an eyelet that is bolted to the intake manifold.

■ The wire connectors at the charge temperature sensor and at the detonation (knock) sensor.

■ The wire connector at each fuel injector.

■ The three-way connector at the distributor.

■ The connector at the O_2 sensor.

■ The connector at the vehicle speed sensor.

■ The two connectors at the power module.

■ The connector at the CTS.

In checking connectors and connections, establish first if they are loose. If one becomes suspect, remove the connector from the terminal and look inside the receptacle. If it is a connection, such as a ground cable bolted to the engine, look beneath the connecting point for contamination. See that connectors are not damaged and there is no dirt or corrosion.

If a connector had a layer of grease, apply a thin layer of dielectric grease to one-half of the connector before you reconnect it. If there is

Fig. 15-3. Only those electrical connectors treated with dielectric grease during production should be treated with dielectric grease when servicing the vehicle. If a connector has grease already applied to it, apply more. If there is no grease, don't apply any.

no grease in the connector, clean the connector, but do not apply grease. Make sure the terminal to which the connector attaches is clean (Fig. 15-3). Use force in reattaching connectors—they must be secure.

RELEASING FUEL SYSTEM PRESSURE

Once you determine that an electrical fault is not causing the engine performance problem you're encountering, turn your attention to the mechanical aspects of MPFI troubleshooting. Troubleshooting begins by making the fuel system pressure test. Be sure your pressure gauge is calibrated to record at least 10 psi above the pressure at which the system is rated. For instance, in the case of the MPFI system for the four-cylinder engine equipped with turbocharger, that rating is 53 psi. Therefore, a pressure gauge calibrated to record at least 63 psi is necessary.

Before doing the fuel system pressure test, it is necessary to bleed pressure from the fuel system. If you don't do this, gas will spray as you attach the fuel pressure gauge, creating a fire hazard. Chrysler recommends that you follow these steps to release fuel system pressure:

1. Remove the gas tank cap.

2. Disconnect the wire connector from a fuel injector.

3. Take a jumper wire outfitted at each end with an alligator clip and connect one alligator clip to either of the two fuel injector terminals. Then attach the other alligator clip to a metallic part of the engine, such as a bolt head or bracket. This grounds the injector.

4. Now connect the alligator clip on the end of a second jumper wire to the other terminal of the fuel injector. Extend the jumper to the positive terminal of the car battery and touch the clip on the end of the jumper to that terminal for no longer than 10 seconds. This action releases fuel system pressure.

5. Disconnect both jumper wires and put the gas tank cap back in place. You can now proceed safely to do the fuel system pressure test.

Caution: If there is any hint of a gas leak, such as an odor or fluid around a fitting, do not release pressure using the method just outlined. Instead raise the vehicle and pull apart the fuel pump connector lying just outside the fuel tank. Then crank the engine, which will probably start and run for a few seconds. When the engine stops running, pressure has been released. Be sure to reconnect the fuel pump connector. Doing so is necessary in order to make the fuel system pressure test.

FUEL SYSTEM PRESSURE TEST

The fuel system pressure test is done by disconnecting the fuel delivery line coming from the fuel tank at the point where it connects to the fuel rail. The fuel rail holds the fuel injectors. Connect the fuel pressure gauge between the end of the fuel delivery line and the fuel rail.

Now turn on the ignition switch and allow the fuel pump to pump gas for a few seconds before you crank the engine. Then start the engine and allow it to run for several seconds before noting the reading on the fuel pressure gauge. This specification varies from engine to engine. If a shop manual is not available, check with a Chrysler dealer for the correct specification for your vehicle. With the four-cylinder engine equipped with a turbocharger, for example, the reading on the fuel gauge should be 53 psi, plus or minus 2 psi.

If the gauge records the correct pressure, there is nothing mechanically wrong with the system. Shut off the engine, release fuel system pressure, disconnect the fuel pressure gauge, reattach the fuel delivery line to the fuel rail, start the engine, and check for fuel leaks.

On the other hand, if fuel pressure is less than that called for by the specification, release fuel system pressure, disconnect the fuel pressure gauge, and reattach the fuel delivery line to the fuel rail. Then disconnect the fuel delivery line where it enters the in-line fuel filter and attach the fuel pressure gauge at this point. Turn on the ignition.

If the fuel gauge now reads the specified pressure, the in-line fuel filter is clogged and should be replaced. Before you disconnect the fuel pressure gauge to do this, make certain you again release fuel system pressure.

Suppose, though, that the fuel gauge shows a reading that is no different from the one displayed when you did the test at the fuel delivery line–fuel rail connection. Do this:

Start the engine. As it idles, gently squeeze the fuel return line coming off the FPR. One of two things will happen:

1. If fuel system pressure as recorded by the pressure gauge rises, the FPR is damaged. Replace it.

2. If fuel system pressure remains below specification, the problem causing the letdown in performance lies either with the fuel sock or fuel pump. In either instance, the fuel pump assembly has to be removed from the fuel tank.

There is an additional possibility that may occur when you perform the fuel system pressure test with the gauge connected between the fuel delivery line and fuel rail. You could get a pressure reading that is substantially in excess of the specification. The reason is either a damaged FPR or a clogged or kinked fuel return hose.

To determine which of these problems exists, turn off the engine, release fuel system pressure, and pull the fuel return line off its fitting on the FPR. Attach a length of substitute hose to that fitting, putting the unattached end into a container. Keep the fuel pressure gauge connected.

Start the engine. If the fuel pressure gauge now shows that pressure within the fuel delivery system is to the specified reading, the fuel return line is restricted. Inspect it to see whether it is kinked. If it is, straighten the line and do the test again. But if test results prove to be the same, replace the fuel-return hose.

Replacing parts of the Chrysler MPFI system is similar to replacing the same parts in GM and Ford MPFI systems. If you aren't familiar with the procedures, refer to Chapters 6, 7, and 10, which cover the removal and installation methods used by these other manufacturers, and apply them to your Chrysler.

However, replacing the fuel injectors is quite a job. The air plenum has to be taken off. The fuel rail can then be unbolted, and the rail and injectors removed as an assembly. Fig-

Fig. 15-4. The FPR is a fairly easy part to replace. It is bolted to the fuel rail. Remove bolts to release it.

ures 15-4 to 15-8 demonstrate what's involved in replacing some major parts associated with the multiport fuel-injection system on a Chrysler 3.0-liter V-6 engine.

Fig. 15-5. To get at elements of the Chrysler MPFI system that are part of the engine, relieve fuel pressure and unbolt the air plenum cover.

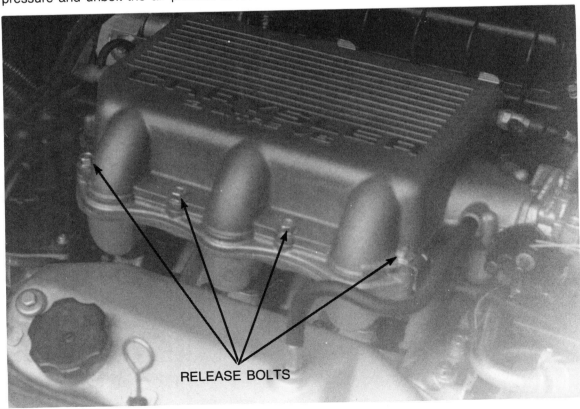

RELEASE BOLTS

Fig. 15-6. Lift the air plenum cover off the manifold. If the cover is cracked, replace it.

Fig. 15-7. Inspect air manifold pipes that extend into the engine for contamination. Be sure that the throttle valve (arrow) moves freely.

Fig. 15-8. The fuel injectors and fuel rail are an assembly. However, if any fuel injector is faulty, it can be released from the fuel rail by undoing the retaining clip (arrow) and pulling the injector out of the engine.

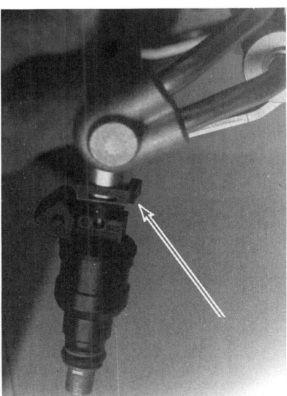

16 Foreign Car Electronic Fuel Injection Systems

How They Work

MOST FOREIGN CAR MANUFACTURERS, among them Toyota and Nissan, use an EFI system based on a design developed by the Robert Bosch Company of West Germany. The system is designated the L-Jetronic system and is computer controlled. The L part of the term stands for *Luft,* which is the German word for air. This refers to the air flow meter, which is one of the components involved in the system's operation. The term Jetronic has no particular meaning, but is used as a fancy catchword for its impact on potential customers.

The L-Jetronic system is a multiport EFI system. Using data suppled by the air flow meter and the ignition system, an electronic control unit (ECU) is able to control the duration for which fuel injectors serving the cylinders of the engine stay open (Fig. 16-1). The air flow meter monitors the rate of air flow into the engine. The ignition system monitors engine speed.

THE GAS AND AIR CIRCUITS

Gas and air take separate routes into the engine.

In the gas circuit, gas flows to the engine from the gas tank. It passes through a fuel filter (called a sock) on the fuel pickup end of an electric fuel pump, the fuel pump, a fuel pulsation damper, through another fuel filter that is outside the gas tank, and through a fuel delivery line or pipe to four, six, or eight fuel injectors implanted in the engine. Each injector sprays gas at the intake valve of a cylinder. There are four fuel injectors if the engine has four cylinders, six if it possesses six cylinders, and eight if there are eight cylinders.

The air circuit starts at an air housing that contains a filter to trap dirt carried by air before it is drawn into the circuit. The cleaned air passes through a duct, into and through the

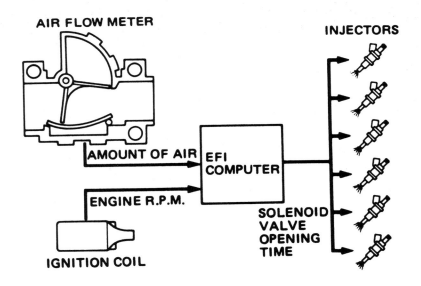

Fig. 16-1. At the heart of the L-Jetronic EFI system is the computer, which transposes data from the air flow meter and ignition system into action. Based on information supplied by the air flow meter and ignition system, the computer controls the fuel injectors so they stay open for the period of time necessary for sound engine performance.

air flow meter, and into a throttle body. The throttle body houses a throttle valve, which is controlled via cable as the driver presses the accelerator pedal. Depending upon the opening angle of the throttle valve, a certain amount of air in the throttle body flows past it into the intake manifold, where it mixes with injected gas. The mixture then passes through intake valves into the cylinders.

An understanding of each of these gas and air circuits components—what they do and how they do it—will help you pinpoint a malfunction. To discuss these components, versions of the L-Jetronic EFI system employed by Nissan and Toyota are used as examples. L-Jetronic EFI systems in other foreign cars are similar.

Incidentally, the first-ever Nissan Motor Corporation car sold in the United States to have the L-Jetronic EFI system was the 1975 Datsun 280Z built for sale in California. Since then, most Nissan/Datsun models manufactured for sale in the United States have had a multiport EFI system based on the L-Jetronic concept. (Nissan dropped the Datsun name in the mid-1980s.)

The first Toyota models to possess the L-Jetronic EFI system were the 1981 Supra and Cressida. As with Nissan, the role of this system has been expanded by Toyota to include most models since that year.

Note: From one model to the next, there are variations in the makeup of the system used by these and other car manufacturers.

THE FUEL SOCK AND FUEL PUMP

The fuel pump in a Nissan or Toyota vehicle equipped with the L-Jetronic EFI system is an electrically operated unit that works when the engine is being cranked and is running (Fig. 16-2). This type of electric fuel pump is often referred to as a wet pump, because there is gasoline inside it all the time. Gasoline keeps the pump cool and lubricated. When the engine is not running, a check valve in the pump closes to keep gas from draining out of the pump and back into the gas tank. If gas drained out of the pump, you would have to crank the engine for an abnormally long period to get the engine to start. Long cranking would be needed to prime the pump before the pump was able to begin delivering gas through the fuel circuit.

The electric fuel pump used by Nissan and Toyota possesses an armature that consists of a series of rollers called vanes. As the armature rotates, the vanes kick gas out of the pump through an outlet into the fuel circuit.

In addition to the check valve, the L-Jetronic EFI system fuel pump has a pressure relief valve. This valve closes the outlet port if pressure in the fuel circuit gets too high. The max-

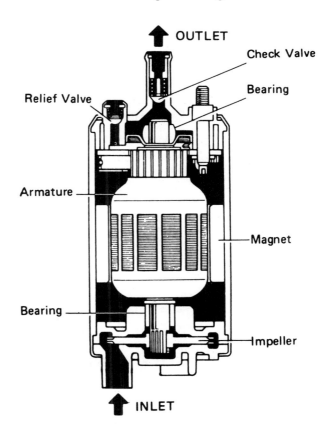

Fig. 16-2. The electric fuel pump of the L-Jetronic EFI system used by Toyota and Nissan is a rotary vane unit that has pressure relief and check valves.

imum amount of pressure the fuel system can handle without causing a disruption in engine performance is 43 to 64 psi, depending upon the model of the vehicle. If pressure exceeds the limit, the engine will flood and be hard to start, will stall, or will lose power. With the pressure relief valve sealing the outlet port, gas keeps circulating inside the pump until pressure drops to a normal level. At this point, the pressure relief valve opens and gas again flows through the outlet port into the fuel circuit.

The sock, which is a coarse-mesh filter, is at the fuel pickup end of the fuel pump. Its job is to prevent large particles of dirt that may be in gasoline from getting into the fuel pump and the rest of the fuel circuit. This contamination would damage components.

The sock is self-cleaning and usually does not have to be replaced. Foreign matter is washed off the filter by gas and falls harmlessly to the bottom of the fuel tank.

FUEL PULSATION DAMPER AND FUEL FILTER

As gas is pumped by the fuel pump from the tank into the fuel circuit, it goes through a fuel pulsation damper (Fig. 16-3). The fuel pulsation damper contains a spring-loaded diaphragm which has the job of absorbing surges that result as gas is being pumped. Without the fuel pulsation damper, these surges would be felt by those in the car. The fuel pulsation damper also muffles any noise that these surges might make.

When gas exits the fuel pulsation damper, it travels to an in-line fuel filter (Fig. 16-4). The job of this fine-mesh filter is to trap small particles of dirt carried by gasoline, which the sock is not designed to trap. If dirt gets inside a fuel injector, it will clog the injector and impede the delivery of gas. Subsequently, the engine would run badly.

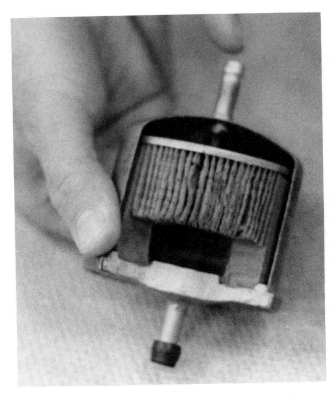

Fig. 16-4. The in-line fuel filters of Toyota and Nissan L-Jetronic EFI systems, as well as those of all other vehicles possessing the L-Jetronic EFI system, have fine-mesh filtering elements that trap impurities carried by gas. Replace this filter at the interval suggested by the vehicle's manufacturer.

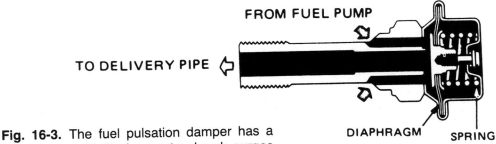

Fig. 16-3. The fuel pulsation damper has a spring-activated diaphragm to absorb surges in fuel pressure as gas is pumped from the fuel pump into the fuel circuit.

The manufacturer of your car may recommend that you replace the in-line fuel filter on a scheduled basis. For example, Nissan recommends replacing this filter every 25,000 miles if the car is a 1975 to 1977 model. If it is a newer model, the suggested interval is either every 30,000 miles or as needed, depending on the model. Consult maintenance requirements outlined in your owner's manual or the vehicle's service manual concerning the manufacturer's recommendation regarding how often to replace the in-line fuel filter.

THE FUEL INJECTORS

The fuel injectors of the L-Jetronic EFI system are mounted in the intake manifold and are aimed at the cylinder intake valve ports (Fig. 16-5). Injectors are electrically operated, variably timed solenoid valves, which means that they stay open and spray gas as long as they are energized with electricity.

Depending on the model of your car, all fuel injectors are either energized and de-energized at the same time or are set up to function in groups with an alternating sequence. In those models where simultaneous energizing takes place, all injectors open and close at the same time. In those models employing an alternating energizing sequence, sets of injectors open and close together. In a six-cylinder engine, for example, three of the injectors may be coupled to open simultaneously, while the other three are set to close as a group. Alternate sequencing is often called sequential operation.

The car's ECU—call it the computer if you wish—energizes the fuel injectors. In addition to data it receives from the air flow meter and ignition system, the ECU also receives information from sensors concerning various engine parameters, such as coolant temperature, air temperature, and the oxygen (O_2) content of the exhaust. Based on this information, the ECU is able to control how long fuel injectors remain open to meet the needs of the engine for the specific driving conditions at the time.

The ECU uses voltage from the battery to do this. As a safety feature in using battery voltage, there is a voltage-dropping resistor in the circuit between the ECU and fuel injectors. This resistor is necessary to reduce battery voltage, which at 12 volts is in excess of what injectors can handle. Without the resistor bringing voltage down to a safe level, injectors would burn out.

In pre-1982 Nissan models equipped with the L-Jetronic EFI system, the voltage-dropping resistor is a separate component and is visible and serviceable. In post-1982 Nissans and all Toyotas with the L-Jetronic EFI system, the voltage-dropping resistor is built into the ECU. If it goes bad, the ECU has to be replaced.

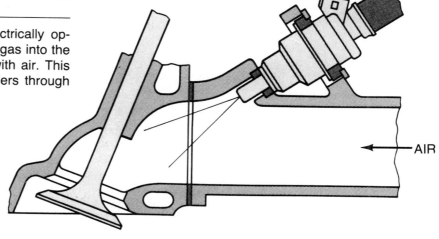

Fig. 16-5. Fuel injectors are electrically operated solenoid valves that spray gas into the intake manifold, where it mixes with air. This mixture then flows into the cylinders through intake valves.

OTHER PARTS ASSOCIATED WITH THE FUEL CIRCUIT

The fuel pressure regulator (FPR) and a cold-start assembly consisting of a cold-start valve, thermotime switch, and an air regulator are other important components that may be part of the L-Jetronic EFI system in your car.

Fuel Pressure Regulator

If not for this part, fuel injectors would spray too much gas into the engine and create an imbalance between fuel system pressure and engine manifold vacuum. By keeping pressure and vacuum in balance, the FPR prevents flooding.

The FPR derives information about the changing state of intake manifold vacuum through a line that extends from it to the intake manifold (Figs. 16-6, 16-7). Based on this information and the pressure of gas flowing into it from the fuel delivery line (fuel rail), the FPR controls a fuel return port to maintain consistent pressure in the fuel system. The FPR opens this port to permit gas to bleed from the fuel circuit into a fuel return line, which returns gas to the fuel tank when an imbalance is imminent. This bleed-off keeps pressure constant. In most Nissan and Toyota models, fuel circuit pressure is maintained at 36 psi, give or take 1 or 2 psi.

Cold-Start System

The L-Jetronic EFI system in your car may have a cold-start system to facilitate starting in cold weather. If a car does not have a cold-start system, the ECU assumes the task performed by the cold-start system. That task is to provide the extra shot of gas a cold engine needs to start promptly. The ECU does this by allowing the fuel injectors to remain open longer when the engine is being cranked.

A cold-start system uses an additional fuel injector called a cold-start valve. Its job is to spray gas into the engine when the engine is being cranked. In addition to the cold-start valve, the cold-start system consists of a

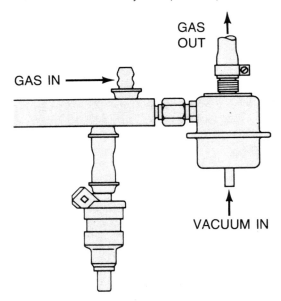

Fig. 16-6. There are three lines attached to the FPR of an L-Jetronic EFI system. Gas enters the FPR through the fuel intake line or pipe. Another line is attached to a port to monitor engine manifold vacuum. A third line is used to send gas back to the fuel tank. These three lines help the FPR to maintain a constant pressure in the fuel circuit.

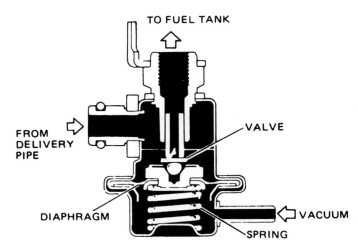

Fig. 16-7. The FPR houses a spring-activated diaphragm that moves in response to engine vacuum, thus allowing gas to enter a return chamber and be delivered back to the fuel tank.

temperature–time switch, which is often called the thermotime switch, and an air regulator. The following is a description of each part:

The *cold-start valve* is mounted just behind the throttle valve and ejects gas directly into the air stream coming into the intake manifold through the air regulator. The cold-start valve, which receives current directly through the cranking circuit of the ignition switch, is energized only when the engine is being cranked. As soon as the ignition key is released, voltage to the cold-start valve ceases and the valve becomes inoperative.

The *thermotime switch* acts as a fail-safe device to prevent flooding if a cold engine doesn't start promptly. The thermotime switch causes the cold-start valve to cease operation if the engine doesn't start promptly and the driver continues to crank the engine, which would cause flooding (Fig. 16-8). At −4° F, for example, the thermotime switch shuts off the cold-start valve if the engine doesn't start within 12 seconds by breaking the electric circuit to the valve.

The thermotime switch has another function. It is installed in the thermostat housing at the front of the engine where it is in contact with engine coolant to monitor coolant temperature. In so doing, it prevents the cold-start valve from operating when the engine is warm.

The *air regulator* of the cold-start system introduces the air that mixes with gas being sprayed into the engine by the cold-start valve. It also lets the engine run at fast idle until it warms up. This higher idling speed is often necessary to prevent a cold engine from stalling.

A heating element inside the air regulator is energized by voltage received from the battery. As the element gets hot, the heat it produces is applied to a bimetallic thermostatic device. As the bimetal gets hotter and hotter, it causes a shutter in the air regulator to close gradually and seal the air passage into the intake manifold. As a result, engine idling speed lowers progressively as the engine warms and fast idling is no longer necessary.

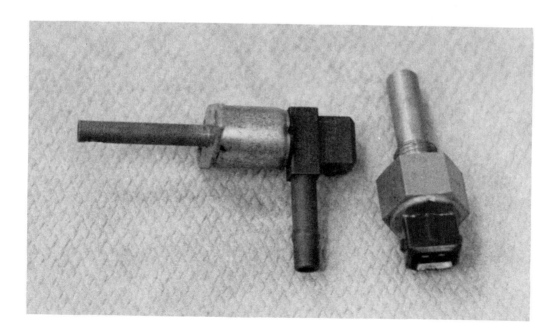

Fig. 16-8. The cold-start valve (left) sprays gasoline into the engine for as long as the thermotime switch (right) allows.

PARTS OF THE AIR CIRCUIT

As mentioned earlier in this chapter, the air circuit of the L-Jetronic system consists of an air cleaner, air flow meter, and throttle valve (Fig. 16-9).

The air flow meter is connected electrically to the ECU and provides the ECU with data in the form of voltage signals about air. A hinged plate that monitors air flow volume is the key part in providing these signals.

In order for air to pass through the air flow meter and subsequently through the throttle valve into the intake manifold, it must push open this hinged plate. The opening angle—5 degrees, 10 degrees, 20 degrees, or whatever—is established by the amount of air that presses against the plate. The volume of air, in turn, is established by the position of the driver's foot on the accelerator pedal, which operates the throttle valve. The more the accelerator pedal is pressed, the more the throttle valve opens. This creates a stronger vacuum pull, which causes the hinged plate to open more.

The particular voltage signal sent to the ECU by the air flow meter varies with the position of the hinged plate. The ECU responds by keeping the fuel injectors open in proportion to the volume of air entering the engine. Thus, the proper air-to-gas mixture ratio needed by the engine to perform properly is maintained.

When the engine runs at idle, the throttle valve is closed and there is no vacuum pull on the hinged plate of the air flow meter. Consequently, no air can get into the engine and the engine could flood. To prevent this from happening, there is a bypass around the air flow meter and throttle valve that allows air to circumvent these closed plates and mix with gas during engine idling periods.

Chapters 17 and 18 discuss how to troubleshoot the mechanical parts of the L-Jetronic EFI system, using Nissan and Toyota as examples. Keep in mind that if testing these parts doesn't reveal the reason for an engine performance problem, the electronic system, which controls fuel injection, should be tested according to procedures outlined in the shop manual for your vehicle.

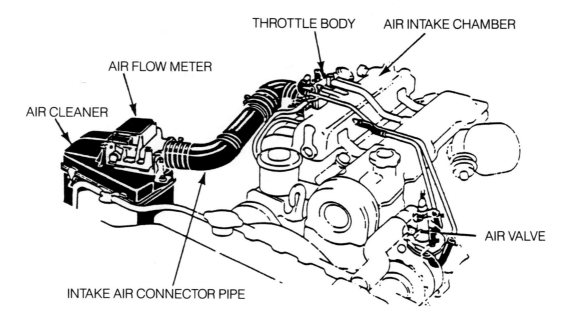

Fig. 16-9. The air intake unit of the L-Jetronic EFI system consists of the air cleaner, air flow meter, and a throttle body containing the throttle valve. The air valve is part of the cold-start assembly.

17

Foreign Car Electronic Fuel Injection Systems

Preliminary Troubleshooting and Repair

Fig. 17-1. A performance problem with an engine possessing an L-Jetronic EFI system is often caused by a loose or leaking hose.

Fig. 17-2. Check all hoses for cracks. Make sure they are securely attached.

IF YOUR FOREIGN CAR develops an engine performance problem that may be caused by a malfunction in the L-Jetronic EFI system, first do the preliminary inspections outlined in this chapter. If this doesn't prove fruitful, make a fuel pressure check. This is also described in this chapter. The final step is to turn to Chapter 18, which outlines a step-by-step troubleshooting procedure to follow for specific performance problems.

FINDING LEAKS

If an engine equipped with an L-Jetronic EFI system that has been performing satisfactorily suddenly falters, the reason is often an air or vacuum leak. A leak will cause the engine to stall, idle roughly, misfire, lose power, hesitate, and/or surge (Fig. 17-1).

The first step in searching for an air or vacuum leak is to inspect all air and vacuum hoses for tears and cracks (Fig. 17-2). The number and locations of hoses vary from car to car. If a service manual for your particular car is not available, be guided by Fig. 17-3, which pinpoints possible areas of leaks in cars using the L-Jetronic EFI system. Pay particular attention to accordion-type hoses (often called ducts). Cracks frequently develop within the folds and are hard to spot.

Loose fittings that hold hoses to components are other spots susceptible to leakage. A can of carburetor cleaner is helpful in checking these areas. Start the engine. While it's running at idling speed, spray the cleaner around one fitting at a time. If engine speed changes, you have located a loose fitting. Tighten the fitting and do the test again.

A damaged engine gasket is another potential source of leakage—for example, a bad valve cover gasket. Look for oil seeping from spots like this. If oil is leaking out of an engine from an opening, then air can leak into the engine through the same opening. Therefore, if an engine drivability problem exists and you spot an oil leak, check out that leak before tearing into the EFI system. To do this, start

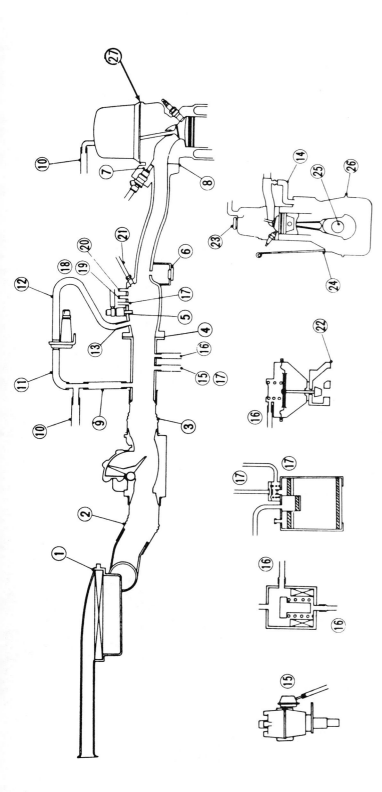

1. Air cleaner element
2. Air duct
3. Air duct
4. Flange (throttle chamber to intake manifold)
5. Cold-start valve mounting surface
6. Blind plug (EGR)
7. Injector mounting surface in intake manifold
8. Cylinder head mounting surface in intake manifold
9. Hose (throttle chamber to 3-way connector), both sides
10. Hose (3-way connector to rocker cover), both sides
11. Hose (3-way connector to air regulator), both sides
12. Hose (air regulator to throttle chamber connector), both sides
13. Throttle chamber connector mounting surface
14. Hose (pipe connector to PCV valve), both sides
15. Distributor vacuum line
16. EGR vacuum line
17. Canister vacuum and purge line
18. Master vacuum line
19. Cooler vacuum line (same vacuum hole as 18)
20. Automatic transmission vacuum line
21. Pressure regulator vacuum line
22. EGR valve mounting surface
23. Oil filler cap
24. Oil level gauge
25. Oil seal (on front and rear of crankshaft)
26. Oil pan gasket mounting surface
27. Valve cover gasket

Fig. 17-3. This composite sketch shows points of possible air and vacuum leaks for all engines using the L-Jetronic EFI system.

the engine. As it runs at idle speed, spread SAE 40 or SAE 50 motor oil over the suspected area. If engine speed changes, a leak is present. Tighten the cover and see whether the engine performance problem has been eliminated. If it hasn't, the gasket may have to be replaced.

LOOKING FOR AN ELECTRICAL MALFUNCTION

Make sure the charging and ignition systems are performing efficiently. If you are not familiar with procedures involved in testing these systems, consult a professional technician (Fig. 17-4).

Once you have checked on the charging and ignition systems, tackle the connectors that tie together wires that deliver current to the EFI system (Fig. 17-5). The number and locations of these connecting points differ from car to car, so consult a service manual for the particular model car.

Check each connector for looseness, bent terminals, and corrosion. Keep in mind that just because a connecting point is tight and looks clean on the outside doesn't mean it is clean on the inside. Pull each connector apart and look inside. If terminals are corroded, clean them with a small brush. If terminals are bent, straighten them with needle-nose pliers.

Fig. 17-4. Electricity begins at the car's charging system. If this is not functioning correctly, the L-Jetronic EFI system will not function correctly. Therefore, the charging system and battery should be tested.

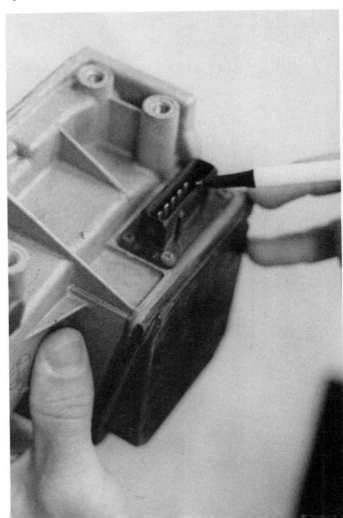

Fig. 17-5. Be sure to examine each electrical connecting point for clean terminals. Also, make certain that wires are connected tightly. The connector illustrated in this photograph is that of the air flow meter of the L-Jetronic EFI system.

Reconnect the two parts of a connector securely.

Note: If terminals had grease on them to begin with, apply grease to them before reconnecting the two parts of the connector. Use dielectric grease, which is a special formulation that allows conductivity. Dielectric grease is available from an auto parts dealer and from stores that sell electronic equipment.

RELEASING FUEL SYSTEM PRESSURE

The reason for doing a fuel system test at this point in the troubleshooting procedure is to establish whether the cause of the performance problem you are having with an engine lies in the fuel circuit of the L-Jetronic EFI system. Before doing the fuel system test, however, release pressure in the system; otherwise, gas will spray all over the place when you disconnect the fuel line, creating a hazardous situation.

To release fuel system pressure, use a service manual to find the fuel pump relay in the engine compartment. This relay is not in the same place from one car to another. Furthermore, there are usually other relays lying alongside the fuel pump relay (Fig. 17-6), which makes identification difficult without the service manual as a reference. However, if you don't have a manual, you can still try to locate the proper relay by using the following procedure:

1. Start the engine and let it run at idling speed.

2. Disconnect the wire harness from a relay. If the engine stalls, you may have hit the fuel pump relay. However, you can't be sure. It could be an ignition system relay. Disconnecting it will cause the engine to stall, but pressure within the fuel system will still be maintained.

3. To make sure the relay you have disconnected is the fuel pump relay, go to the rear of the car, put your ear near the fuel tank, and have someone in the car turn on the ignition. If you don't hear the fuel pump making its characteristic whirring noise, you *have* disconnected the fuel pump re-

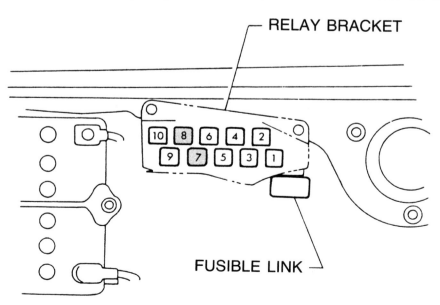

Fig. 17-6. A car equipped with the L-Jetronic EFI system has quite a number of electrical relays. Therefore, finding the one for the fuel pump (number 8 in this case) can be a guessing game.

lay. Crank the engine a few more times to release any residual pressure in the system. Then, reconnect the wire harness to the relay and proceed to do the fuel system pressure test.

Suppose you can't seem to identify the fuel pump relay. Don't give up yet. There's one other possibility that may let you release fuel pressure without use of the relay.

Raise the car and look for a connector on the outside of the gas tank. If there is one, it is probably a fuel pump wire harness connector. Many model cars with the L-Jetronic EFI system have this connector. It ties together a wire from the car's electrical system to a wire that is attached to the fuel pump.

If you find a connector, pull its two parts free of each other. Then, start the engine. If the engine stalls, the fuel pump has indeed been disconnected. Crank the engine a few more times to clear residual pressure, turn off the ignition, reconnect the connector, and proceed with the fuel system pressure test.

TESTING FUEL SYSTEM PRESSURE

You need a fuel pressure gauge to test whether the fuel system in a car with an L-Jetronic EFI system is building up enough pressure to deliver gas to the fuel injectors. Be sure to use a gauge that's been designed to test the particular level of fuel system pressure in your make of car. This level differs from one car to another, so check the specification in the service manual. Fuel pressure gauges are available from the parts department of a dealer selling your make of car and from auto parts stores.

Before you do this next step, look for a test valve on the fuel rail similar to the one used by GM and Ford on their MPFI systems. The fuel rail is the elliptical pipe to which fuel injectors attach and that circumscribes the engine. If there is a valve, you can connect the fuel pressure gauge to it and you don't have to disconnect the fuel line. Most cars with L-Jetronic EFI systems do not have these test valves.

If your car does not have a test valve on the fuel rail, find the fuel delivery line coming from the in-line fuel filter at the point where the line connects to the fuel rail. Disconnect the fuel delivery line from the fuel rail, and connect the fuel pressure gauge between the line and fuel rail. Be sure to tighten connectors to avoid loss of pressure and a gas leak around pressure gauge fittings as you conduct the test (Fig. 17-7).

Make sure the fuel pump relay or fuel pump wire harness connector have been reconnected. Then start the engine and check the pressure gauge as the engine runs at idle. The gauge should record the fuel delivery pressure for the engine as specified in the service manual.

If fuel pressure does not respond in this way, there is a failure somewhere in the fuel system and the delivery of gas is being impeded. How to handle the situation and go about troubleshooting other parts of the L-Jetronic EFI system are described in detail in Chapter 18.

Caution: Before you disconnect the fuel pressure gauge, you again must release fuel system pressure.

Fig. 17-7. Install the fuel pressure gauge between the fuel rail and in-line fuel filter. Be sure connections are tight.

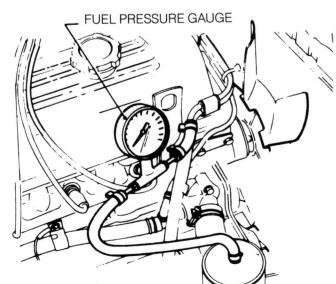

18

Foreign Car Electronic Fuel Injection Systems

A Part-by-Part Check

ONE OF EIGHT engine performance problems will result when a part of your car's L-Jetronic EFI system malfunctions:

1. The engine won't start.
2. The engine will be hard to start.
3. The engine will start and stall.
4. The engine will idle too fast when it is cold.
5. The engine will misfire.
6. The engine will lack power.
7. The engine will hesitate (stumble) as the car is accelerated.
8. The engine will surge.

This chapter describes each of these problems in terms of the EFI components that are most likely to cause them. The chapter also provides step-by-step troubleshooting procedures to follow.

Fig. 18-1. The fuel pump check valve usually can be replaced by removing the pump and unscrewing the valve from the pump housing.

ENGINE WON'T START OR IS HARD TO START

The same malfunction that makes an engine hard to start can lead to complete starting failure if the malfunction is allowed to continue. Therefore, these two performance problems, which are frequently caused by the same malfunctioning part, are discussed together.

Here are the steps, in the order to follow, if an engine with the L-Jetronic EFI system displays a hard starting or no-start problem:

1. Test the battery. It must be capable of providing a minimum of 9.5 volts of cranking power with the engine cold.

2. Check to make sure the ignition system is performing properly and that spark plugs are in sound condition.

3. Verify that the intake and exhaust valves are adjusted to the specifications given in the service manual for your particular engine.

4. If a fuel system pressure test (Chapter 17) confirms that there is a problem in the fuel delivery system, check to see if the fuel pump is working. Stand in the rear of the car with your ear close to the fuel tank so you can hear the pump and have someone turn the ignition switch on without cranking the engine. You should hear a whir or buzzing for several seconds, confirming that the fuel pump is functioning. If you don't hear a noise, do the following:

■ Make sure the fuse protecting the fuel pump circuit hasn't blown by putting a new fuse into the fuse panel.

■ Test the fuel pump electric circuit, including the fuel pump relay.

■ Replace the fuel pump.

Important: If your hard-to-start engine does not respond to tests outlined in this section, take a look at the fuel pump check valve described in Chapter 17. If the valve goes bad, it will cause hard starting. There is no test that will reveal a bad valve. The fuel pump will have to be removed from the car, the valve replaced, and the fuel pump reinstalled to see if the hard-starting problem has been eliminated (Fig. 18-1).

5. Check the operation of the electrical circuit serving the fuel injectors. This can be done with an EFI-LITE as described in Chapter 6, that deals with the GM multipoint fuel injection system (Fig. 18-2). If the test reveals a circuit failure, have an electrical specialist check the ECU power input circuit, ignition coil trigger input circuit, ECU ground circuit, and the circuit between the battery, fuel injectors, and ECU.

6. Check the operation of the fuel injectors. If the electrical circuit serving the fuel injectors is working okay, one or more fuel injectors may be causing the starting problem. You can find out if a fuel injector is not working by letting the engine run and placing the tip of a screwdriver on one injector at a time. Put your ear near the handle of the screwdriver (Fig. 18-3). A clicking sound means the injector is operating.

A more accurate way of checking fuel injectors is with a Pulse Duration Tester, which can be ordered from Kent-Moore Tools, and carries tool number J-3398 (see page 24 for the address). At the time this book was being written, the tool was selling for about $60. Connect the testing instrument according to instructions that accompany the tool.

The Pulse Duration Tester energizes the injector in millisecond increments, so you can determine whether the injector is opening and closing at the intervals specified by the manufacturer for good engine

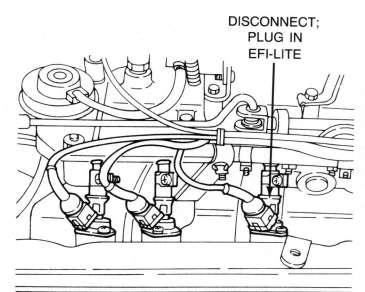

Fig. 18-2. The same procedure described in Chapter 6 for using an EFI-LITE to test for electricity at the fuel injectors of a GM multipoint fuel injection system can be used to test the circuit of an L-Jetronic EFI system.

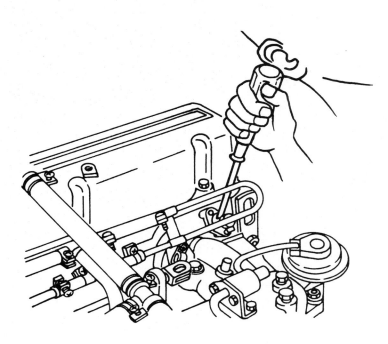

Fig. 18-3. If the fuel injector is working, a clicking will be transmitted by the screwdriver to your ear.

Fig. 18-4. The Pulse Duration Tester displays readings in milliseconds to let you determine whether each injector is supplying too little, too much, or no gas at all.

performance according to prevailing driving conditions (Fig. 18-4). A reading of 0 means the injector is not opening and no gas is getting through. A reading below 4 means the injector is closing too quickly and a lean-fuel condition exists. A reading in the 4 to 7 range signifies normal injector performance. A reading above 7 means the injector is staying open too long and a rich-fuel condition exists.

7. If the engine is hard to start and it has a cold-start system, make certain the cold-start valve is being energized as the engine is cranked. This can be tested with the EFI-LITE. If you don't get power to the valve, trace the cold-start circuit with a 12-volt test light to determine the location of the failure. To do this, find the thermotime switch by tracing the wire back from the cold-start valve to the switch. The thermotime switch has two wires connected to it—one comes from the ignition switch, and the other comes from the cold-start valve (Fig. 18-5).

Disconnect the wire coming from the ignition switch, and attach the test light to its terminal. Connect the other end of the test light securely to a ground connection. Have someone crank the engine. If the test light glows, current is getting to the thermotime switch. Reconnect the wire. If the test light doesn't glow, troubleshoot the circuit from the thermotime switch back to the ignition switch.

Suppose the test light glows, indicating that the thermotime switch is getting current. In this case, disconnect the wire from the other terminal of the switch—the one coming from the cold-start valve. Attach the test light to the terminal of the thermotime switch and to ground. Crank the engine. If the test light glows, current is getting through the thermotime switch, but not to the cold-start valve. Therefore, there is a problem in the circuit between the switch and cold-start valve. But if the test light doesn't glow, there's trouble with the thermotime switch. Replacing it should get the engine to start promptly.

If the circuit from the ignition switch to the cold-start valve is acting normally, the reason for hard starting when the engine is cold could lie with a faulty cold-start valve. Since the cold-start valve is a fuel injector, you can use the Pulse Duration Tester to

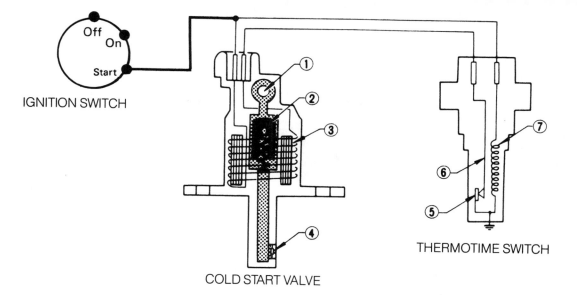

Fig. 18-5. This illustration depicts the electrical circuit between the ignition switch, cold-start valve, and thermotime switch. Numbers indicate the following: (1) fuel inlet of the cold-start valve; (2) cold-start valve internal mechanism; (3) cold-start valve electric coil; (4) fuel outlet of cold-start valve; (5) thermotime switch contact points (circuit is activated with points closed); (6) bimetal thermostat; (7) resistance heater.

find out. However, for the test to be valid, make sure the engine is cold and being cranked as you take the reading.

ENGINE STARTS AND STALLS

A dirty air filter (Fig. 18-6), clogged in-line fuel filter (Fig. 18-7), or malfunctioning fuel injector often causes this problem. Replace filters if necessary. Then test fuel injectors.

Fig. 18-6. To ensure that the flow of air into the fuel circuit is not being hampered—a condition that would cause flooding and stalling—replace a dirty air filter and clean the air filter housing.

Fig. 18-7. A dirty fuel filter will impede the flow of gas to the fuel injectors and thus to the engine. This will cause a lean-fuel condition, which is a primary cause of stalling.

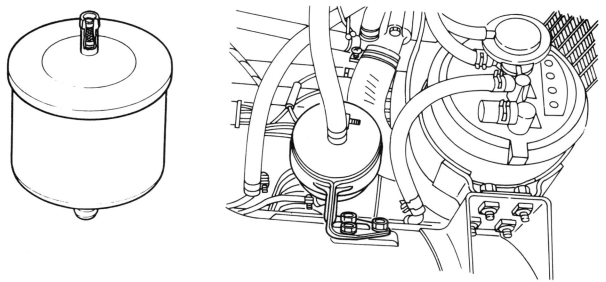

Fig. 18-8. Determine whether the hinged plate of the air flow meter moves freely.

A malfunctioning air flow meter can also cause stalling. Take the air intake hose off the nozzle of the air flow meter and see that the air flow plate moves freely (Fig. 18-8). If it is binding and can't be freed, replace the air flow meter.

ENGINE IDLES TOO FAST

If the engine idles too fast when it is first started, make sure the intake and exhaust valves are adjusted for proper clearance. Then test the functioning of the air regulator, which is also called the air valve (Fig. 18-9).

This part has an air line that bypasses the air flow meter. It allows the delivery of air into the cold-start system. Air mixes with the gas supplied by the cold-start valve, so the engine can start easily and run properly when cold.

To test the air regulator, start the engine (cold) and pinch closed the hose attached to the regulator. If idling speed falls, replace the air regulator. If idling speed continues at its rapid pace, look into the following non-EFI related conditions: manifold vacuum leak, PCV valve malfunction, and a bad seal around the dipstick and oil filler cap.

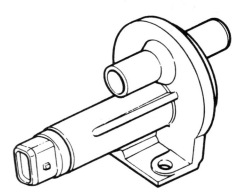

Fig. 18-9. Suspect that the air valve (regulator) of the L-Jetronic EFI system cold-start setup is restricted if the engine idles too fast when it is first started. The valve may have to be replaced.

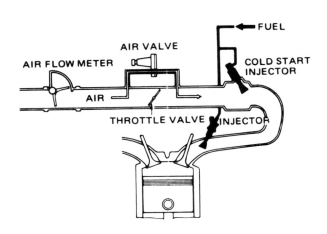

ENGINE MISFIRES

Begin looking for the cause of this problem by making a complete inspection of the ignition system, especially of spark plugs. Another non-EFI condition to investigate is a leaking exhaust gas recirculation (EGR) valve. This can be done by taking the EGR valve off the vehicle, attaching a hand-held vacuum pump to its hose fitting, and pumping up vacuum. Vacuum should hold steady. If the valve has a ruptured diaphragm, the drop in vacuum as shown by the meter of the hand-held vacuum pump gauge will be dramatic. As for EFI-related conditions that can make an engine misfire, they are as follows:

- Corroded or loose EFI harness connectors. Pull each connector apart, look for bent terminals (straighten them with needle-nose pliers), clean off corrosion, apply dielectric lubricant to the terminals if terminals were previously treated, and secure the two parts of the connector together tightly.

- Improper fuel pressure caused by a bad FPR. Release fuel pressure and remove the FPR from the car to test it using a hand-held vacuum pump (Fig. 18-10).

- A fuel delivery problem that is restricting the flow of gas to the engine (Fig. 18-11). This fault most often lies with the in-line fuel filter, FPR (see above), or a fuel injector.

ENGINE LACKS POWER; HESITATES; SURGES

All three problems can be caused by the same malfunction. Check the following parts of the engine until the fault is found:

- Ignition system.

- Air and vacuum hoses, and vacuum components including the PCV and EGR valves.

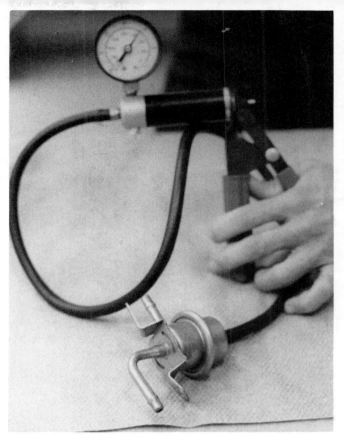

Fig. 18-10. The FPR will fail only if the rubber diaphragm inside the chamber is damaged. Test it by attaching a hand-held vacuum pump to the vacuum fitting. Pump up vacuum. If the gauge doesn't rise and hold steady, install a new FPR.

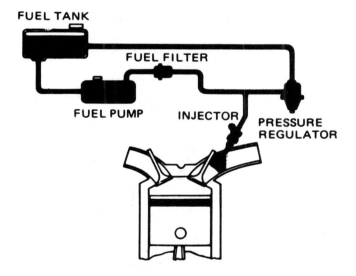

Fig. 18-11. The fuel circuit of the L-Jetronic EFI system is made up of only a few parts, which makes it relatively easy when troubleshooting to find the reason for engine misfire.

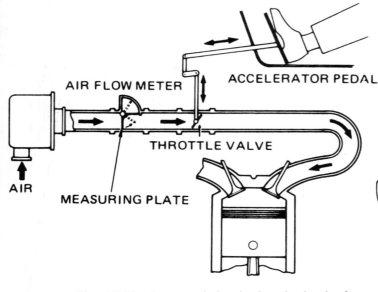

Fig. 18-12. Any restriction in the air circuit of an L-Jetronic EFI system will result in an overly rich fuel mixture. An overly rich fuel mixture will cause faulty engine performance.

- Air cleaner element. If dirty, replace it.

- Hinged front plate of the air flow meter. Move the plate by hand to make sure it does not bind.

- Throttle valve. Make certain the connections between the accelerator pedal and throttle valve are clean and tight. Any impediment can cause a sluggish throttle valve and a reduction of air that has to mix with gas for a proper fuel mixture (Fig. 18-12).

- Lack of fuel pressure. Do a fuel pressure test as described in Chapter 17.

- Trouble with one of the sensors feeding data to the ECU. Have the air temperature

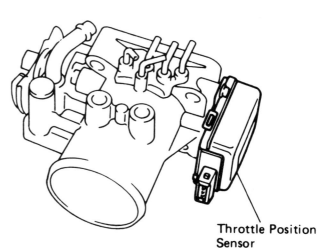

Fig. 18-13. The TPS of the L-Jetronic EFI system is on the throttle body. It senses the angle of the throttle valve opening and transmits the signal to the ECU. Using data from this and other sensors, the ECU determines whether to increase or decrease the period of time fuel injectors stay open.

sensor, CTS, O_2, sensor, exhaust and TPS tested according to instructions provided in your car's service manual (Figs. 18-13 to 18-16). Sensors are parts of the vehicle's electronic control system and are used by the ECU to determine how long the fuel injectors should remain open. If a sensor fails, the ECU is not able to determine the particular parameter monitored by the sensor. Figure 18-17 illustrates a complete L-Jetronic EFI system to include the gas circuit, air circuit, and ECU.

Fig. 18-14. One type of TPS that has been used in the L-Jetronic EFI system has sets of points that open and close in relation to the opening angle of the throttle valve. If the points stick, the signal sent to the ECU will not be correct. This, in turn, will result in too much or too little gas being ejected into the engine. Poor performance will be the outcome.

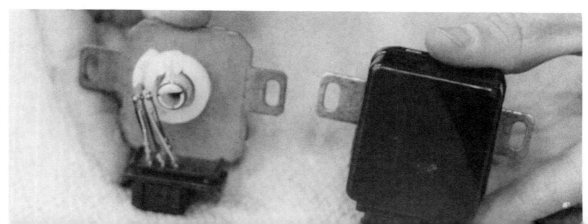

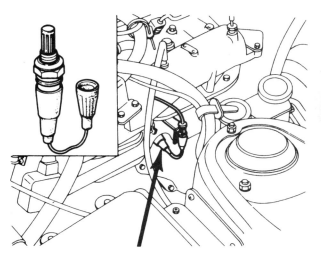

Fig. 18-15. Another important sensor is the exhaust gas sensor, which is also called the O_2 sensor, because it monitors the density of the oxygen in the exhaust. Using the information gathered by the O_2 sensor, the ECU is able to adjust the amount of gas to keep it in line with the amount of air. The result is the ideal mixture the engine needs to run properly.

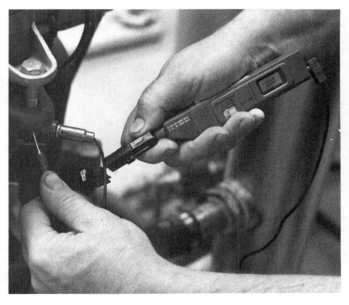

Fig. 18-16. The O_2 sensor is tested with a VOM. Readings should be as specified in the service manual. If they aren't, replace the sensor.

Fig. 18-17. The complete L-Jetronic EFI system is portrayed in this illustration to include the gas circuit, air circuit, and computer-controlled circuit.

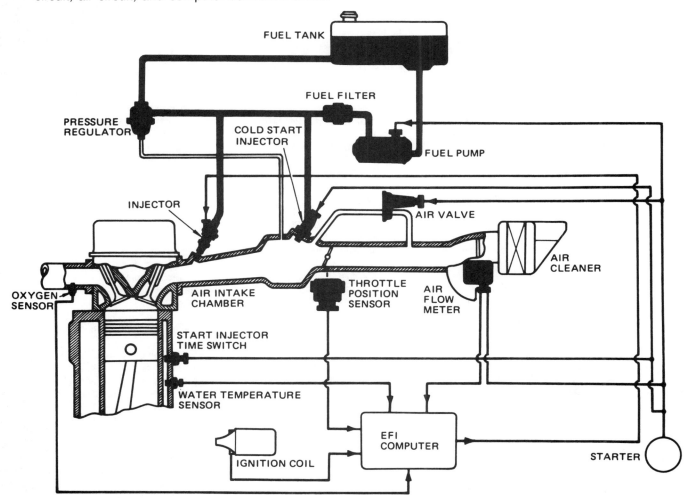

Index

AC/Delco X66-P, 49
air bypass valves, 3, 95
air circuits:
   Chrysler multiport fuel injection system, 130–131
   components, 148
   failure, 59–60
   Ford central fuel injection systems, 66
   foreign car electronic fuel injection systems, 142–143, 148, 150–152
   GM multiport fuel injection system, 44, 46, 59–60
   GM throttle body fuel injection system, 19–20
   leaks, 150–152
air cleaner ducts, loose or cracked, 75
air cleaner elements, dirty, 162
air cleaner housing, checking, 113
air cleaners, 148
   vacuum motors, 3
air conditioning wide-open throttle cutout relay trouble code, 115
air density, 124–125
air filter elements, replacing, 110
air filters:
   clogged, 9
   dirty, 159
   replacing, 59
air flow meters, 142–143, 148
   hinged front plates, 162
   malfunctioning, 160
air horns, 66
air intake systems, *see* air circuits
air regulators, 147
   rapid idle, 160
air temperature sensors, 162
automatic idle speed (AIS) motors, 104, 131
   Chrysler single-point electronic fuel injection system, 109
   trouble code, 115
   troubleshooting, 120–122
automatic shutdown (ASD) circuits, 102
   trouble code, 115
   wiring connectors, 114

batteries, 156–157
black exhaust smoke, 40
Borroughs Tool and Equipment Corporation, 24, 26, 49
Robert Bosch Company, 142
Buick 3.8-liter V-6 engines, 8

carburetor systems, vii, 12
   feedback, 14
charcoal canister purge solenoid trouble code, 115
charcoal canister purge valves, 3, 5
charcoal canister purge valve vacuum switches, 112
charge temperature sensors, 125–126
   electrical connectors, 135
charging system, trouble code, 115
CHECK ENGINE SOON light, 16, 38, 114
check valves, failing, 56
Chrysler multiport fuel injection (MPFI) system, 123–139, vii
   air circuits, 130–131
   description, 124
   electrical systems, 124–126, 135–136
   electrical system troubleshooting, 135–136
   FPR, 127–129, 137
   fuel delivery system, 126–131, 136–137
   fuel filters, 129–130, 137
   fuel injectors, 130, 135
   fuel pressure, 126–129
   fuel pressure testing, 137
   overview, 123–131
   relieving fuel delivery system pressure, 136
   throttle body components, 131
   troubleshooting and repair, 133–139
Chryslers:
   2.2-liter four-cylinder engines, 108
   2.5-liter four-cylinder engines, 108
   3.0-liter six-cylinder engines, 130
   3.0-liter V-6 engines, 108
   5.2-liter V-8 engines, 108
   5.9-liter V-8 engines, 108

four-cylinder engines, 108, 135–136
Chrysler single-point electronic fuel injection (EFI) system, 99–122
   AIS motor, 109
   AIS motor troubleshooting, 120–122
   computers, 101–102
   description, 101
   electronic considerations, 99–108
   FPR, 109, 117
   FPR replacement, 120
   fuel delivery system, 101
   fuel delivery system components, 109–110
   fuel filter replacement, 119
   fuel filters, 110
   fuel injector replacement, 119–120
   fuel injectors, 108–109
   fuel injector wiring connectors, 114
   fuel pump, 109–110
   fuel system pressure testing, 116–117
   makeup of system, 105–110
   relieving fuel delivery system pressure, 115–116
   self-diagnosis, 114–115
   sensors, 102–104
   TBI part replacement, 119–120
   throttle body, 106–109
   trouble codes, 114–115
   troubleshooting and repairing, 111–122
   troubleshooting problems, 115–116
cold-start system, 146–147
cold-start valves, 146–147
   engine no-start and hard starting problems, 158
compression:
   leaks, 9
   tests, 8
computer command control (C3), 14–16
   malfunctions, 16
continuity tests, 81
coolant temperature sensors (CTSs), 95, 103–104, 125–126, 162
   electrical connectors, 135
   GM throttle body fuel injection

Index

systems, 20
trouble code, 115
curb idling speed:
   Ford 3.8-liter engines, 75–78
   Ford 5.0-liter engines, 78–79

Datsun 280Z, 143
deceleration valves, 3
detonation (knock) sensors
   electrical connectors, 95, 135
DEXRON II, 36
diagnostic pressure test valves, 79–80
diagnostic pressure valves, 91
   Ford multipoint fuel injection systems, 89
dielectric grease, 153
dieseling, 88
   stuck fuel injectors, 35
distributor circuit:
   electrical connectors, 135
   trouble code, 114
distributor modulator valve assemblies, 3
distributor vacuum advance, 3
double-wrench technique, 52

EFI-LITE, 24, 46, 115, 135, 157–158
electrical systems:
   Chrysler multiport fuel injection system, 124–126, 135–136
   failure, 24–25, 46–47
   faulty connectors, 9–10
   Ford central fuel injection systems, 70
   foreign car electronic fuel injection systems, 152–153
   GM multiport fuel injection system, 46–47
   GM throttle body fuel injection systems, 24–25
   malfunctioning, 152–153
   troubleshooting, 135–136
   visual inspection, 135–136
electric bimetal thermostat, inoperative, 72–74
electronic control assembly (ECA), 70
electronic control module (ECM), 40
   air circuit failure, 59–60
   fuel pump failure, 55
   pumping gas, 19–20
   sensors, 15–16, 19
electronic control units (ECUs), 142, 145
   air flow meter connection, 148
   cold starting, 146
   engine no-start and hard starting problems, 157–158
electronic fuel injection (EFI):
   electrical connectors, 152–153
   misunderstanding, vii

non-vacuum-related defects, 9–10
prevalence, vii
types, vii
vacuum-related problems, 3–8
emissions systems, computer-regulated, 16, 101
engine die-out, fuel injector failure, 130
engine no-start problems:
   dirty fuel injectors, 47–48
   electrical failure, 24–25, 46–47
   foreign car electronic fuel injection systems, 156–159
   fuel delivery system pressure testing, 26–28, 80
   low fuel pressure reading, 29–31
   zero fuel pressure reading, 28–29
engine speed, 124
exhaust emissions standards, 14
exhaust gas recirculation (EGR) valves, 3, 5, 95–96
   checking, 112
   engine misfiring, 161
   trouble code, 115
   wiring connectors, 114
exhaust heat control valves, 3
exhaust systems, restricted, 9
exhaust valves, 156

fan relay trouble code, 115
fan temperature sensors, 95
fast idle speed, Ford 5.0-liter engines, 78
feedback (electronic) carburetor systems, 14
fire extinguishers, 52
flooding:
   causes, 147
   FPR failure, 41
Ford central fuel injection (CFI) systems, 61–83
   air circuit, 66
   air cleaner duct loose or cracked, 75
   description, 63–65
   electrical systems, 70
   electric bimetal thermostat inoperative, 72–74
   FPR, 68–69
   fuel delivery system, 66–69
   fuel delivery system pressure testing, 80–83
   fuel delivery system servicing, 79–80
   fuel injectors, 82–83
   fuel pump inertia switches, 80–81
   fuel pumps, 82–83
   high-pressure, 65–66, 69
   idling speed incorrect, 75–79
   leaking gasket between fuel charging assembly and intake

manifold, 74–75
low-pressure, 65–69
malfunction chart, 73
overview, 61–70
troubleshooting and repair, 71–83
Ford EFI Pressure Gauge, 80–83, 91
Ford multipoint fuel injection (MPFI) systems, vii
   description, 87–88
   diagnostic pressure valves, 89
   FPR, 91
   FPR replacement, 93–95
   fuel delivery system leaks, 89
   fuel delivery system testing, 90–91
   fuel filter replacement, 92–93
   fuel injector replacement, 95–96
   fuel injectors, 91
   fuel pumps, 88, 91
   sensors, 87–88
   TPS removal, 97–98
   troubleshooting and repair, 85–98
   troubleshooting preparation, 88–89
   troubleshooting procedure, 89–92
Fords:
   2.3-liter turbocharged engines, 95–96
   3.8-liter engines, 75–78
   5.0-liter engines, 72–74, 78–79
   Capri, 78–79
   fuel filter replacement, 92–93
   fuel injector replacement, 95–96
   Mustang, 78–79
   troubleshooting and repairing, 72
   troubleshooting preparation, 88–89
foreign car electronic fuel injection (EFI) systems, 141–163
   air circuits, 142–143, 148, 150–152
   cold-start system, 146–147
   description, 142
   electrical connectors, 152–153
   electrical system malfunctions, 152–153
   engine misfiring, 161
   engine no-start problems, 156–159
   FPR, 146, 161
   fuel delivery system, 142–147, 153–154, 156
   fuel delivery system pressure lack, 162
   fuel delivery system pressure testing, 154
   fuel filters, 145
   fuel injectors, 145, 157–159
   fuel pulsation dampers, 144
   fuel pumps, 143–144, 156–157
   hard starting, 156–159
   how they work, 141–148
   part-by-part check, 155–163
   performance problems, 156–163
   power lacks, hesitation, and surging, 161–162

preliminary troubleshooting and repair, 149–154
rapid idle, 160
relieving fuel delivery system pressure, 153–154
sock filters, 144
stalling, 159–160
vacuum and air leaks, 150–152
fuel charging assemblies, 63
   air circuit, 66
   fuel delivery system pressure testing, 81
   leaking gaskets, 74–75
fuel delivery systems:
   Chrysler multiport fuel injection system, 126–131, 136–137
   Chrysler single-point electronic fuel injection system, 101, 109–110, 115–117
   components, 109–110, 126–129
   computer-regulated, 101
   engine no-start and hard starting, 156–157
   Ford central fuel injection systems, 66–69, 79–80
   Ford multipoint fuel injection systems, 89–98
   foreign car electronic fuel injection systems, 142–147, 153–154, 156, 157, 162
   GM multiport fuel injection system, 44, 46, 53–58
   GM throttle body fuel injection systems, 23–31
   high pressure reading, 31
   leaks, 89
   low pressure reading, 29–31
   normally operating, 81
   pressure lacks, 162
   pressure testing, 25–31, 80–83, 116–117, 137, 154
   relieving pressure, 25–26, 53, 80, 115–116, 136, 153–154
   servicing, 79–80
   testing, 90–91
   troubleshooting and repairing, 23–31
   zero pressure reading, 28–29
fuel diagnostic valves, 81
fuel economy, poor, 40
fuel efficiency standards, 14
fuel filters:
   blocked, 82
   Chrysler multiport fuel injection system, 129–130, 137
   Chrysler single-point electronic fuel injection (EFI) system, 110, 117, 119
   clogged, 91, 137, 159
   Ford central fuel injection systems, 68

Ford multipoint fuel injection systems, 92–93
foreign car electronic fuel injection systems, 145
fuel delivery systems pressure testing, 26, 28–31
GM multiport fuel injection system, 44, 54
replacing, 54, 92–93, 119
role, 129–130
fuel hoses:
   clogged, 137
   kinked, 137
   pressure testing, 117
   specifications, 109
fuel injector cleaner, 47
fuel injectors:
   balance testing, 47–48
   Chrysler multiport fuel injection system, 130, 135
   Chrysler single-point electronic fuel injection system, 108–109, 114, 119–120
   dirty, 47–48
   electrical connectors, 114, 135
   engine no-start and hard starting problems, 157–158
   force-cleaning, 49–50
   Ford central fuel injection systems, 66, 82–83
   Ford multipoint fuel injection systems, 91, 95–96
   foreign car electronic fuel injection systems, 145, 157–159
   GM multiport fuel injection system, 44, 46–50, 57–58
   GM throttle body fuel injection systems, 35–38
   inspecting, 35–38
   malfunctioning, 82–83, 91, 159
   nozzles, 37
   replacing, 95–96, 119–120
   role, 130
   stuck, 35, 57–58
fuel lines:
   blocked, 82, 91
   Ford central fuel injection systems, 68
   fuel delivery systems pressure testing, 26–27, 31, 90–91
   GM throttle body fuel injection systems, 19
   high pressure readings, 31
fuel meter assemblies, removal, 35–39, 41
fuel pressure regulators (FPRs):
   Chrysler multiport fuel injection system, 127–129, 137
   Chrysler single-point electronic fuel injection (EFI) system, 109, 117, 120

damaged, 137
engine misfiring, 161
Ford central fuel injection systems, 68–69, 82–83
Ford multipoint fuel injection systems, 91, 93–95
foreign car electronic fuel injection systems, 146
fuel delivery system pressure testing, 117
GM multiport fuel injection system, 45, 56–57
GM throttle body fuel injection systems, 19
malfunctioning, 82–83, 91
replacing, 41, 93–95, 120
testing, 56–57
fuel pulsation dampers, 144
fuel pump check valves, 157
fuel pump relays, 26, 29
   damaged, 35
   inspecting, 35
fuel pumps, 26
   Chrysler multiport fuel injection system, 126–127
   Chrysler single-point electronic fuel injection system, 109–110
   engine no-start and hard starting problems, 156–157
   failure, 54–55
   faulty, 91
   Ford central fuel injection systems, 66–68, 82–83
   Ford multipoint fuel injection systems, 88, 90–91
   foreign car electronic fuel injection systems, 143–144, 156–157
   GM multiport fuel injection system, 54–55
   hearing test, 54–55
   high-pressure, 88
   inertia switches, 80–81
   low-pressure, 88
   malfunctioning, 82–83
   pressure capabilities, 126–127
fuel socks, 44, 54
   fuel delivery system pressure testing, 26, 28–31, 90–91
fuel system pressure gauges, 26–27

gas circuits, *see* fuel delivery systems
GM multiport fuel injection (MPFI) system, vii, 43–60
   air circuit, 44, 46
   air circuit failure, 59–60
   check valve failure, 56
   Diagnostic Tester, 47
   dirty fuel injectors, 47–48
   electrical failure, 46–47
   force-cleaning injectors, 49–50

Index

FPR, 44
FPR failure, 56–57
fuel delivery system, 44, 46, 53–58
fuel filter replacement, 54
fuel injectors, 44, 46–50
fuel pump failure, 54–55
hard starting causes, 56–58
preliminary troubleshooting and repair, 43–50
relieving fuel system pressure, 53
stuck fuel injectors, 57–58
troubleshooting and repairing other components, 51–60
troubleshooting precautions, 52
GM throttle body fuel injection (TBI) systems, 11–41, vii
  air circuits, 19–20
  computer command control, 14–16
  coolant sensors, 20
  cross-fire, 13–14
  description, 12–13
  ECM, 15–16, 19–20
  electrical failure, 24–25
  FPR, 19
  FPR replacement, 41
  fuel delivery system, 25–31
  fuel delivery system pressure testing, 26–28
  fuel injector inspection, 35–38
  fuel lines, 19
  high fuel pressure reading, 31
  how they work, 17–20
  IAC failure, 38–40
  IACs, 20
  ISC failure, 38
  ISCs, 20
  low fuel pressure reading, 29–31
  MAP sensors, 20
  model 700, 36
  non-EFI part roles, 20
  O_2 sensors, 20
  overview, 11–16
  sensors, 15–16, 19
  single-point, 13–14
  stoichiometric ratio, 14
  throttle body servicing, 33–41
  TPS removal, 40
  troubleshooting and repairing fuel delivery system components, 23–31
  two-by-one (2 × 1), 13–14
  two-point (dual), 13–14
  varieties, 12–13
  zero fuel pressure reading, 28–29
GM Top Engine Cleaner, 49

hard starting, 73, 74, 88
  bad check valves, 56
  causes, 56–58
  dirty fuel injectors, 47

foreign car electronic fuel injection systems, 156–159
  FPR failure, 56–57
  inoperative electric bimetal thermostat, 74
  stuck fuel injectors, 57–58
head gaskets, leaking, 9
hesitation, 45, 73, 88, 130
  dirty fuel injectors, 47
  foreign car electronic fuel injection systems, 161–162
  inoperative electric bimetal thermostat, 74
  TPS failure, 40
high fuel pressure readings, 31

idle air controls (IACs), 20
  GM multiport fuel injection system, 59–60
  replacing, 38–40
idle speed control (ISC) motors, 20, 77–78
  replacing, 38
idling speed, 160
  Ford 3.8-liter engines, 75–78
  Ford 5.0-liter engines, 78–79
  incorrect, 75–79
ignition systems, 142, 156
  computer controlled, 16, 101
  engine misfiring, 161
  power lacks, hesitation, and surging, 161
  timing, 9
injection on-time (pulse width), 19
injector wiring harnesses, 95
intake manifolds, leaking, 74–75
intake valves, 156
in-tank fuel delivery assemblies, 31
internal relief valves, 66–68

Kent-Moore Tool Group, 24, 26, 49, 157
knock (detonation) sensors, 95, 135

leaking gaskets, 74–75
lean-fuel symptoms, 130
limp-in mode, 126
Lincolns:
  fuel filter replacement, 92–93
  troubleshooting and repairing, 72
  troubleshooting preparation, 88–89
L-Jetronic systems, see foreign car electronic fuel injection systems
  logic module, 124
  air circuit role, 131
  fuel injector control, 130
  trouble code, 115
low fuel pressure readings, 29–31

manifold absolute pressure (MAP) sensors, 3, 102, 125

checking, 112
electrical system trouble code, 114
GM throttle body fuel injection systems, 20
vacuum system trouble code, 114
wiring connectors, 114
manifold vacuum, 8
  leaks, 160
Mercurys:
  fuel filter replacement, 92–93
  troubleshooting and repairing, 72
  troubleshooting preparation, 88–89
metal fuel pipe replacement, 52
misfiring, 161
multiport fuel injection (MPFI) systems (multipoint fuel injection systems), vii
  components, 44
  other names, 44
  vacuum hose routing, 5

Nissans, 142–143
  voltage-dropping resistors, 145
non-vacuum-related defects, 9–10
nozzles, fuel injector, 37

ohmmeters, fuel pump testing, 90
optical distributors, 124
O ring seals, 52
overlapping valves, 8
oxygen (O_2) sensors, 6, 103, 162
  electrical connectors, 135
  GM throttle body fuel injection systems, 20
  trouble code, 115
  wiring connectors, 114
ported vacuum, 8
positive crankcase ventilation (PCV) valves, 3, 95
  checking, 113
  faulty, 160
power lack, 73, 88
  foreign car electronic fuel injection systems, 161–162
power module, 124
  electrical connectors, 135
  fuel injector control, 130
pressure relief valves, 127
Pulse Duration Tester, 157–159
pulse-time-modulated injection systems, 65; see also Ford central fuel injection systems
purge solenoids, 114

rapid idle, 160
reference pickups, 124
rough idle, 45, 73, 88
  dirty fuel injectors, 47
  IAC failure, 38
  inoperative electric bimetal thermostat, 74

Index

ISC failure, 38
TPS failure, 40
rpm-sensing (reference) pickups, 126
rubberized fuel hose replacement, 52

saddle bracket locking screws, 78–79
self-test connectors, 77
self-test input connectors, 77
sensors:
    Chrysler single-point electronic fuel injection system, 102–104
    Ford multipoint fuel injection systems, 87–88
    GM throttle body fuel injection systems, 15–16, 19
    malfunctioning, 162
sequential fuel injection (SFI), 130
silicone grease, 95
single module electronic control (SMEC), 101–102, 124
sock filters, 54, 110
    foreign car electronic fuel injection systems, 144
    malfunctioning, 83
    replacing, 92
    role, 129
solenoid valves, 46–47
spark plugs, 156
spark timing, 124
speed-density fuel injection systems, 124
speed sensors, 104
stalling, 45, 73, 88, 131
    air circuit failure, 59
    foreign car electronic fuel injection systems, 159–160
    fuel injector failure, 130
    IAC failure, 38–40
    inoperative electric bimetal thermostat, 74
    ISC failure, 38
    preventing, 147
    TPS failure, 40

stoichiometric ratios, 14
stumbling, 45, 73, 88, 130
    dirty fuel injectors, 47
    inoperative electric bimetal thermostat, 74
surges, 45, 73
    foreign car electronic fuel injection systems, 161–162
synchronization pickups, 124

temperature-time switches, 147
thermotime switches, 147
throttle bodies, 130–131
    AIS motor component, 109
    Chrysler single-point electronic fuel injection system, 106–109
    components, 108–109, 131
    electrical connectors, 135
    FPR component, 109
    fuel injector component, 108–109
    servicing, 33–41
throttle body fuel injection (TBI) systems, vii
    replacing parts, 119–120
    vacuum hose routing, 5
    see also GM throttle body fuel injection systems
throttle body-to-manifold joints, 8
throttle modulators, 3
throttle position sensors (TPSs), 95, 103, 125–126, 131, 162
    Ford multipoint fuel injection systems, 97–98
    removal, 40, 97–98
    trouble code, 115
throttle stop adjustment screws, 77
throttle valves, 148
    GM throttle body fuel injection systems, 20
    impeded, 162
Toyotas, 142–143
    Cressida, 143
    Supra, 143
voltage-dropping resistors, 145

vacuum:
    facts about, 3
    leaks, 3–4
    manifold, 8
    ported, 8
vacuum gauges, 7–9
    calibration, 7
    gasket leak checking, 74–75
vacuum hoses, 4–5
    checking, 112–113
    power lacks, hesitation, and surging, 161
    routing, 5
vacuum pumps, using, 3–9
vacuum systems:
    components, 3
    diaphragms, 9
    leaks, 150–152, 160
    sounds and sights, 4–5
    specifications, 8
    testing, 7–8
    VOM O_2 sensor troubleshooting, 6–7, 9
valve guides, worn, 9
vehicle emission control information (VECI) decals, 77–79
vehicle speed sensors:
    electrical connectors, 135
    trouble code, 114
voltage-dropping resistors, 145
voltmeters:
    fuel delivery system pressure testing, 80–81
    fuel pump testing, 90
volt/ohmmeters (VOMs), 6

wet pumps, 143
wire harnesses, checking connectors, 114

zero fuel pressure readings, 28–29